Navigating Innovation

Benoit Gailly

Navigating Innovation

How to Identify, Prioritize and Capture
Opportunities for Strategic Success

Benoit Gailly
Louvain School of Management
Université catholique de Louvain
Louvain-La-Neuve, Belgium

ISBN 978-3-030-08394-6   ISBN 978-3-319-77191-5   (eBook)
https://doi.org/10.1007/978-3-319-77191-5

Cover design: Tom Howey

Printed on acid-free paper

This Palgrave Macmillan imprint is published by the registered company Springer International Publishing AG part of Springer Nature.
The registered company address is: Gewerbestrasse 11, 6330 Cham, Switzerland

Preface

This manager's guide is a one-stop shop for the fundamentals of innovation management best practices that you should master if you want to successfully manage innovation. It can be used by individual managers or students who want to understand innovation management, by teams that want to identify and address specific innovation issues in their project or organization, by executives who want to equip their colleagues with relevant and actionable innovation management insights or by teachers and coaches who are looking for a reference book that is both practice oriented and theory driven.

They can use this guide in a traditional way, to face one innovation challenge after another, or jump directly to key issues, insights or specific knowledge nuggets in order to find the content they need.

Innovative organizations need to act like entrepreneurs, identifying, prioritizing and exploiting opportunities. They must also manage opportunities in line with their strategy and develop the organization and ecosystem required to take advantage of them. This guide is therefore structured along those five challenges:

- Introduction: Make sense of innovation
- Challenge 1: Build a shared *strategic* vision of innovation
- Challenge 2: Manage entrepreneurial *ecosystems*
- Challenge 3: *Identify* attractive innovation opportunities
- Challenge 4: Develop a balanced *portfolio* of business models
- Challenge 5: Nimble *execution*: fail fast and win big
- Conclusion: More brain, less storming

In order to allow the reader to easily navigate these chapters, each one is organized around a set of core issues and key insights (see figure), and each chapter concludes with a synthesis of those key insights and selected references. An extended bibliography for each of the core issues is also provided in the companion website www.NavigatingInnovation.org.

Innovation challenges

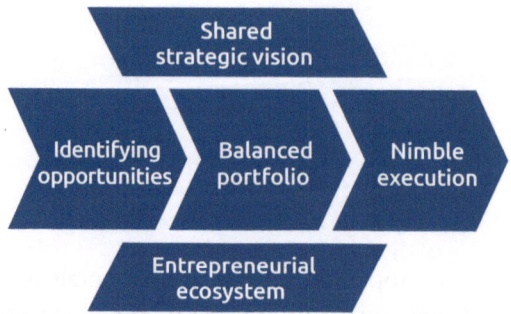

Core issues (Chapter 2 example)

2.1 Why it matters: innovation management capabilities
2.2 What innovation means from a business point of view
2.3 How innovation unfolds over time as a process
2.4 What are the different types of innovation firms can engage in
2.5 What are the strategic innovation options
2.6 What should drive strategic decision-making regarding innovation
2.7 Synthesis

Each key insight is covered as a "knowledge nugget", including a clear, jargon-free explanation illustrated with multiple concrete examples and direct managerial implications.

This guide is therefore not a book of magical solutions and universal innovation management recipes (beware of those!). Nor is it an in-depth scientific literature review of all innovation issues and concepts (there are entire journals, collections and libraries devoted to just that). But it is the right one-stop shop to start mastering the fundamentals of innovation management.

Louvain-La-Neuve, Belgium   Benoit Gailly

Acknowledgements

This guide is based upon nearly 20 years of experience training and coaching master's students and executives, participating in research projects and academic conferences with leading international scholars, and supporting firms in the design and development of successful innovation management capabilities. The ideas, insights and examples gathered here would not exist without fruitful exchanges and collaborations with many innovation apprentices and champions.

Since 2001 I have also benefited from interactions with hundreds of international masters and executive students and from the wisdom and findings of the experts and scholars listed in the bibliography of this book. I am particularly grateful for what I learned from discussions with, as well as writings and presentations by, my colleagues Profs. Norbert Alter, Olivier Basso, Paul Belleflamme, John Bessant, Raymond Collard, Nathalie Delobbe, Françoise de Viron, Alain Fayolle, Xavier Pavie, Bernard Surlemont, Armin Schwienbacher, Ludo van der Heyden and Olivier Witmeur.

But innovation knowledge and insights are relevant only if they can be put into practice and help managers cope with innovation challenges. I am therefore grateful to all the innovation champions I had the opportunity to work with, in particular (in alphabetical order):

Hugues Bultot, CEO, Univercells
Frédéric Burguet, Deputy Manager, Strategic Innovation–Vehicle Information
    Technology
Mireille Buydens, Lawyer and Partner, Janson Baugniet
Eric Cornuel, Director General and CEO, EFMD
David Dab, Chief Innovation Officer, ING Belgium

François Cornelis, former Executive Vice President of Total

Luc de Brabandere, author and former Partner, Boston Consulting Group

Benjamin Dessy, Vice President Digital Payment & Labs, Mastercard

Benoit Domercq, New Business Development Manager, AGC Glass Europe

Leopold Demiddeleer, former Director Future Business, Solvay

Vincent Duprez, Senior Vice President Innovation, Safran Aero Boosters

Christophe Hilbring, Senior Product Manager, Mastercard

Peter Hinssen, Technology Entrepreneur, Lecturer and Author

Yves Jongen, Chief Research Officer and Founder, IBA

Hugues Langer, Chief Technology Officer, Sonaca Group

Brigitte Laurent, former Group Innovation Champion—SVP, Solvay

Philippe Lemmens, former CEO Belgacom Skynet and Partner, Leanstudio

Philippe Mauchard, co-founder, McKinsey Solutions

John Metselaer, former Innovation Center Leader, P&G and Director, Innovation Council

Dominique Neerinck, VP Technology and R&D at Total Gas, Renewables and Power

Lorna Payne, Managing Director, OneLife

Frédéric Sallmann, Head of Innovation at GSK Vaccines

Jean-Marie Solvay, Member of the Board of Directors, Solvay

Jean Stephenne, former President, GSK Biologicals

Luc Sterckx, former CEO, Luminus

Grégoire Talbot, founder and former CEO, Cockpit Group

Jean-Yves Tilquin, Group R&D Director, Carmeuse Group

Pierre Tossut, Group R&D Director and Channels Director, Puratos

Marc van den Neste, CTO Building & Industrial, AGC Asahi Glass

Sven Vandeputte, Managing Director, OCAS

Guido Vandervorst, Managing Partner Innovation, Deloitte

Luc Vansteenkiste, former CEO, Recticel

Michel Vlasselaer, Partner, Roland Berger Strategy Consultants

Vincent Werbroeck, former Chief Innovation Officer, Maggotteaux

Last but not least, a special thank-you to Roland Nellissen and Bernard Querton ("Kanar"), who helped me prove once again that some pictures are worth a thousand words.

Contents

List of Figures

1

Introduction: Make Sense of Innovation

Innovation is today on top of the agenda of managers, entrepreneurs and policymakers. It is an opportunity to develop and grow new business models but also a threat to existing ones. It is also a societal challenge, as entire professions disappear while new ones are created. Simply ignoring innovation is therefore not an option.

But innovation debates too often seem more like buzzword competitions than rigorous management thinking. From digital transformation to industry 4.0, from sustainable business models to crowd-hackathons, managers can get lost in what has often become a confusing maze of ideas, concepts, tools and initiatives.

As a consequence, too many managers, entrepreneurs and policymakers still live and think in "innovation wonderland", a place where great opportunities are just an ideation workshop away, where spending more on R&D or creativity sessions is the key to success, where crowds always have wisdom and where being the first to invent is always the winning option.

But in the real world of innovation, opportunities must be hunted and matured, distinctive innovation management capabilities must be developed and tough strategic choices regarding innovations must be made. Being a very creative business is one thing; successfully managing innovations is another.

The key to innovation success is to do much more than generate ideas. The key to innovation success is to have an organization capable of effectively identifying, prioritizing and capturing innovation opportunities, in line with its strategy and ecosystem. The key to innovation success is therefore more "brain" and less "storming".

© The Author(s) 2018
B. Gailly, *Navigating Innovation*, https://doi.org/10.1007/978-3-319-77191-5_1

Combining rigorous concepts and hands-on experience, this guide will help you find your way through the innovation maze. It will help you ask the right questions and equip you with the right approaches and insights, thus empowering you to find the answers that will work for *your* organization.

These questions, approaches and insights are organized according to the five key challenges of successful innovation management:

- Build a shared strategic vision of innovation
- Manage entrepreneurial ecosystems
- Identify attractive innovation opportunities
- Develop a balanced portfolio of business models
- Nimble execution: fail fast and win big

Welcome to the real world of innovation. Fasten your seat belt and enjoy the journey!

2

Build a Shared Strategic Vision of Innovation

Innovation is today seen by many organizations as a key strategic issue. These organizations, however, often embark on costly innovation initiatives, without having any clear idea of what they want to achieve (thus they are inefficient) or why (thus they are ineffective). Too often, they do not understand or do not agree on what innovation is or why and how they want to innovate.

The first challenge of innovation management is therefore to develop a shared strategic vision of innovation: why a firm must proactively manage innovation, what innovation actually means as a business, how it unfolds as a process, what are the different types of innovation, what are the resulting strategic options and what should drive the decision-making regarding those options (Fig. 2.1).

2.1 Why It Matters: Innovation Management Capabilities

Innovations have occurred and endured for thousands of years. Fire, boats, pottery, irrigation, agriculture, wheels, writing and dozens of other technological breakthroughs have changed humanity. In recent centuries, major waves of innovation such as mechanization, steam, steel and railroads, electricity and the combustion engine, petrochemicals and electronics have led to industrial revolutions. Each time, organizations, institutions and societies had to learn how to adapt to them.

© The Author(s) 2018
B. Gailly, *Navigating Innovation*, https://doi.org/10.1007/978-3-319-77191-5_2

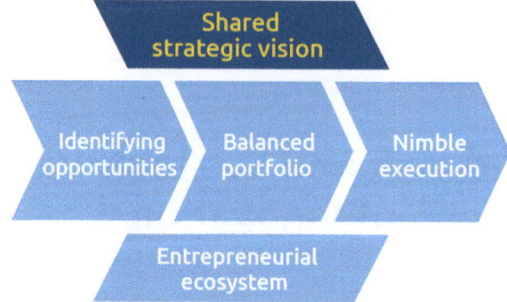

Fig. 2.1 Building a shared strategic vision

What has changed in recent decades is that innovation has evolved from an "exceptional event" to "the new normal". In the past, most firms could cope with innovation as an occasional shock. They enjoyed periods of reasonably stable equilibrium between those shocks. Today, global and industry trends force firms to constantly make decisions regarding existing and emerging innovations. Innovation has to be on the strategic agenda of every firm, and the ability to deal with innovation has become critical (Fig. 2.2).

Key Insights (detailed in following sections)

i. *Megatrends* such as technology disruptions, international competition and sustainability affect firms across all sectors and industries, creating new competitive and social challenges.

ii. *Industries and sectors* are also disrupted by new regulations, new customer needs and new technologies, forcing firms to reconsider the sustainability of their assets and activities.

iii. Small and large firms across sectors must place innovation among their *strategic priorities* if they do not want to suffer the fate of the dinosaurs.

iv. This implies developing *innovation management capabilities* to identify, select and capture the right innovation opportunities, in line with the firm's ecosystem and strategy.

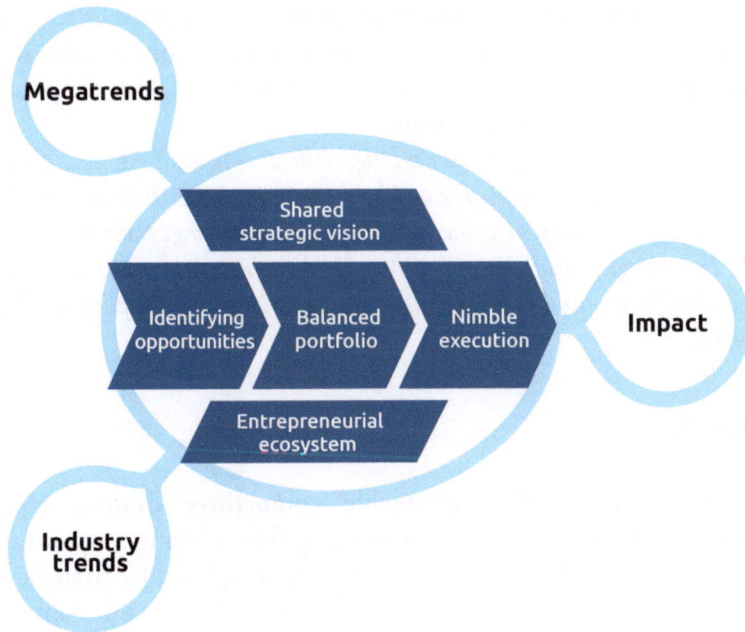

Fig. 2.2   Developing innovation management capabilities

2.1.1 Megatrends: Beyond Hi-Tech

While in the past only the so-called hi-tech sectors were considered innovation intensive, today firms across all sectors are affected by global trends that force them to place innovation on their strategic agenda. Today, managers (still) try to build "immortal firms in mortal markets" (N. Dew) and many employees want lifelong job security, but most firms are bound to suffer the consequences of creative destruction.

Among the global firms that were in the world's top ten (in terms of valuation) in 2000, only one (Exxon) was worth more in 2015 than in 2000. Decades-old household names such as Kodak, Black & Decker, Blackberry and General Motors have disappeared, struggled or faced bankruptcy. One of the oldest firms in the world, Kongo Gumi, a Japanese hotel that opened in 576 and survived countless earthquakes, revolutions and wars, went bankrupt in 2006.

High-Speed Innovations

There are three global innovation-related trends that affect firms across industries. The first one is the accelerating pace of technological development.

While technologies have evolved since prehistoric humans made the first tools, the speed at which new technologies now revolutionize entire sectors is unprecedented. The "innovation crisis" mode is no longer a temporary disequilibrium; it is the new equilibrium.

ICT technologies (new software and algorithms, new datasets, new devices and infrastructures) force all firms to consider how to improve their existing business model, whether to develop new ones, and how to best manage their data and information. Technological innovations such as blockchains, new sensors, new material, drones or additive manufacturing also raise questions across traditional industry boundaries.

The New Normal

As a consequence, building "sustainable" competitive advantage is increasingly the exception, replaced by an ongoing flow of temporary advantages. Strategic planning, with increasing frequency, is replaced by strategic agility, in a world which has become "VUCA" (volatile, uncertain, complex and ambiguous). Again, while this "industrial revolution" is not the first, its pace and intensity are probably unprecedented.

In factories, connectivity, agility, flexibility and recyclability are increasingly replacing unit cost and speed as key performance indicators. In parallel, more and more employees, particularly those in routine and/or manual jobs, have to acquire and develop new skills based on intelligence and/or interaction.

Global Village

The second global trend affecting firms across industries is the internationalization of markets and competition. This "globalization" is not new: the Silk Road has been used for more than 2000 years and European countries have exploited the resources of the whole planet for centuries. Indeed, history teaches us that globalization is not a smooth process without detours and hurdles, as economic integration is often in conflict with democracy and sovereignty. But today even firms in sectors once considered "local", such as food, personal services or entertainment, cannot consider their home market safe and thus have to look beyond their city or region. They can capture opportunities in new geographical areas but also be attacked by competitors they have never even heard of.

Incumbent American and European firms in the automotive, industrial and electronics sectors have been attacked and sometimes overwhelmed by new entrants from so-called emerging countries such as Huawei. While initially focused on low-cost copycat strategies, these new entrants have often become fierce competitors and aggressive innovators.

There Is No Planet B

The third global trend affecting firms across industries is the growing acknowledgement that the impact of consumers and firms on climate, biodiversity and the planet's capacity to sustain aging and future generations must be addressed. This acknowledgement probably began in the 1970s and the first pictures of earth as a small dot lost in space, when the first "green" movements were born and the term "Anthropocene", the epoch when human activity began to exert a lasting ecological impact on Earth, entered informal parlance.

Drought and famine have become frequent causes of conflicts. Exceptional temperatures, floods and storms are becoming the norm. The so-called rare-earth elements are scarce yet most smartphones, batteries and lighting cannot be produced without them. Ten billion people will soon want the facilities and comforts that to date only several hundred million have enjoyed.

So What?

Managers from all sectors should understand how revolutionary technologies, international competition and sustainability are affecting their business model today and will affect it tomorrow. They should also develop a shared understanding of the resulting threats and opportunities, for both their firm and their industry. Faced with these trends, "burying one's head in the sand" and not deciding is actually a decision, and often not a good one.

2.1.2 Industry and Sector Trends

While technology revolutions, globalization and sustainability are global trends affecting firms across sectors, there are also microtrends, creating threats and opportunities in specific industries or sectors. Those microtrends affect in particular what a firm sees as its main current or potential suppliers, competitors, substitutes, partners and customers.

The Future Is Not What It Used to Be

The first microtrend which may affect a sector or an industry is the evolution of the socioeconomic environment in which it operates. New social and cultural evolutions can affect what is accepted or not as legitimate or effective ways of doing business. New policies and/or regulation can affect what are feasible or profitable business models. Macro- or microeconomic developments can affect key profitability drivers such as wages, interest or exchange rates. Finally the diffusion of specific knowledge, technologies or infrastructure can facilitate or hinder existing or new business activities.

In particular, regulations, which are too often considered as always creating a negative burden for firms, can in many cases create opportunities to innovate, by developing new regulatory-compliant business models, by preventing competitive hold-ups or by decreasing existing barriers to entry.

The profitability and success of innovative peer-to-peer and "sharing economy" business models is strongly affected by the cultural, legal, social and technology environment in which they operate. Innovations such as video streaming or online dating are actually quite old and succeeded only when the right socioeconomic conditions emerged. Finally many US-based Internet champions have struggled or failed to compete in the Chinese socioeconomic environment, and conversely.

Fighting for the Cake

The second microtrend which may affect a sector or an industry is the changing balance of power between the different players competing for the value created within that industry or sector. Following Michael Porter's famous model, suppliers try to corner scarce assets or consolidate, customers become more price-sensitive or more versatile, new entrants find ways to avoid barriers to entry or scale economies disadvantages and finally substitutes break incumbents' strongholds. At the same time existing competitors all fight for margins and market shares and new potential partners emerge within and across industries. This evolution of competitive dynamics requires firms to constantly consider whether to continue to defend their existing business models or to develop new ones.

Artist, media companies, publishers, consumer electronics firms and telecommunication networks have all seen their respective profitability radically change as new technologies disrupted the dynamics of the entertainment industry. Similarly, who makes the most money out of the car industry in the future could be energy utilities, specialized equipment suppliers, original equipment manufacturers (OEMs), software providers or online platforms.

New Needs

The third microtrend which may affect a sector or an industry is the evolution of the needs and preferences of its customers and stakeholders. On the one hand the relative importance of different segments of customers and stakeholders can evolve, for example, for social or demographic reasons. On the other hand the expectations of those customers and stakeholders can evolve, for example, in terms of ethics, safety or national preferences. Each of those changes can be an opportunity to innovate or a threat for those who don't.

Financial institutions have seen new segments emerge, such as expatriates, active pensioners or single-parent families. They also sometimes face shareholders for which ethics, local economic development or risks matter much more than in the past. In some countries many consumers now care about environmental, social or health issues they completely ignored just a decade ago.

New Technologies

The last microtrend which may affect a sector or an industry is the diffusion and adoption of new technologies which can improve or disrupt their value chain. Technologies which improve the cost, speed or efficiency of a value chain are a key source of competitive innovation. But technologies which threaten to disrupt industry value chains can also destroy incumbents.

New advances in biotechnologies, data analytics and nanotechnologies can make existing medical treatments more efficient, but also completely change the business models of the pharmaceutical industry, for example, through the development of personalized medicine approaches.

So What?

Managers in each sector or industry should carefully identify and analyze how specific changes in socioeconomic and competitive environment, in markets and in technologies might affect the profitability and sustainability of their business model. They should also understand what are the resulting threats and opportunities, and develop scenarios and contingency plans both for their firm and for their industry.

2.1.3 Innovation as a Strategic Priority

The emergence of global and industry-specific trends means that innovation should now be, for all firms, a strategic issue. It is strategic because there is no obvious optimal choice and because it will affect their future. Each firm must therefore decide how they will cope with these developments, and know that not deciding is implicitly a decision.

A key strategic decision resulting from the analysis of the above trends concerns the level of required innovation and change, which will be a function of the uncertainty and impact of the evolutions in question. The strategic question should therefore not be "whether the firm will innovate", but "how the firm will manage evolutions and innovations".

Safe Heavens (for Now)

If the global and industry-specific evolutions in question are relatively certain to occur, and if the firm perceives that its core assets and activities are not going to be threatened, then it could decide not to change. In this case, it can focus on pursuing its current activities and progressively implement the incremental innovations required to remain competitive.

Retailers dealing with global warming issues can often focus on their existing business models, adjusting only their energy sourcing, product range and waste management approaches. Similarly, airlines dealing with globalization issues can focus on improving their operations, adjusting their pricing policy, schedules and destinations as required.

Icebergs Ahead

If the evolutions considered are relatively certain and threaten the firm's core assets or activities, the strategic priority should be creating a sense of urgency and launching a corporate restructuring. This means identifying the best new business models available, if any, and effectively managing the transition to those new business models (or exit).

The expected increase in CO_2-related taxes and in oil and gas exploration costs is prompting several oil companies to consider downstream and/or alternative energy activities. Similarly, most car companies are investing more in electric batteries and in software and connectivity solutions.

Limited Visibility

If, as is increasingly the case, the evolutions are uncertain, the strategic priority should be to build the technology and market intelligence capabilities needed to minimize the related uncertainty, in order to be "less wrong than the others". This might include investing in customer intimacy, developing learning-by-doing approaches such as rapid prototyping or design thinking, and hedging bets through parallel investment in competing technologies.

Most large financial institutions are carefully following, influencing and investing in the development of new electronic currencies and new payment and investment platforms, and monitoring how they might affect their existing business models.

Embedded Flexibility

On top of this ability to detect and evaluate potential threats and opportunities, firms facing uncertain evolutions that potentially threaten their assets or activities must develop the ability to quickly and efficiently launch and scale up new business opportunities. This means not only recognizing the threats but also developing the strategic vision and capabilities to deal with them.

Nokia recognized as early as 2001 the opportunities to develop personalized online promotions and retail solutions, and to take a share of any online transactions. It failed, however, to develop the capability to capture those opportunities.

So What?

Innovation is a key strategic issue. It does not mean that being the most innovative company is always a winning strategy, or that innovation should become the first priority for all firms, in all sectors. But innovation should be on the strategic agenda. All firms in all sectors should understand which global and specific trends they face. They must then make the strategic choice of which trends they want to lead, which they want to follow, and which they can currently afford to ignore and/or just watch. As the famous management scholar Michael Porter has said, "Strategy is about choosing what not to do."

2.1.4 Developing Innovation Management Capabilities

The challenge for firms is not to innovate all the time and as much as possible but to be able to manage the innovation threats and opportunities that affect the ways they do business and create value. This means developing the entrepreneurial capabilities to identify, evaluate and capture innovation opportunities, as well as the organizational and strategic capability to support and integrate those opportunities.

Corporate champions like Exxon or J&J did not thrive because they were always the most innovative firms on earth. They thrived as long as they were the best at effectively managing the innovations relevant to their strategy.

The Innovation Engine

The first challenge is therefore to build an "innovation engine". This means developing the capabilities to scan the firm's resources and environment in order to identify innovation opportunities, prioritize the most attractive and relevant ones, and to effectively and quickly capture them.

Identifying innovation opportunities involves fostering not only ideation and creativity but also R&D, technology, market intelligence and knowledge management, thus enabling the firm to continuously "sense" its environment.

Prioritizing innovation opportunities involves business model design, selection and prioritization processes as well as portfolio management, empowering the firm to learn and "seize" the right opportunities.

Finally, capturing innovation opportunities involves the business processes and structures needed to implement, accelerate and scale up innovation projects, and in some cases transform the firm.

"Number of new ideas" has never been a key factor in corporate success; being very effective at selecting and implementing a few (sometimes not really) new ideas is.

The Steering Wheel

The second challenge is to make sure that the organization is steering its innovation engine. This means understanding that innovation is not an end, per se, but a means to achieve the firm's strategic objectives. The first question to ask should therefore be "Why do you want to innovate?", then make sure the answers are shared across the organization. This means not only having an innovation strategy but also defining the place of innovation in the firm's strategy.

What is striking about Apple under Steve Jobs is not only what Apple did but also what it chose not to do (such as choosing not to launch a large number of new products) and the focus of his strategy during his tenure as CEO.

The Innovation Fuel

The last (but definitely not least) challenge is to make sure that there is enough "fuel" in the innovation engine. And the key fuel of innovation is people. This means developing the structure and systems but also fostering a culture and climate where corporate entrepreneurs and their teams not only want to but can launch innovative initiatives, often bridging corporate silos and sometimes bending corporate rules. It also means leveraging the resources and talent available in the firm's ecosystem, crossing traditional corporate and industry boundaries.

The biggest challenge for the managers of promising firms and for new business developers is mobilizing talent, not lack of ideas.

So What?

Innovation management is not about ideation or R&D management. It is about building and maintaining the capabilities to effectively identify, select and capture innovation opportunities, in line with the firm's strategy and ecosystem.

2.2　Innovation as a Business: More Than Creativity

The lightbulb metaphor of innovation is probably the worst one ever. It leads too many managers, entrepreneurs and policymakers to believe that when innovation happens, it is both instantaneously triggered (on/off) and immediately diffused (everyone is "enlightened").

As a consequence, organizations too often believe that the best way to manage innovation is to launch brainstorming sessions and ideation processes, and to go "fishing for good ideas". But anyone who has been involved in such

Fig. 2.3 Innovation as a business: more than creativity

processes knows that managing innovation means much more than generating new ideas (Fig. 2.3).

Key Insights

i. Innovation means *much more* than invention. Managing innovation means managing both newness and change, and the latter often matters the most.
ii. *Newness is relative.* What is today new to one manager, its organization or its environment might not be to another.
iii. Innovation is about changing people's perceptions and realities, combining *many small steps and a few big bets.*

2.2.1 Innovation Means Much More Than Invention

Innovation can be defined as the combination of newness and change, or as a change toward something new. An innovation can therefore be defined according to its scope ("what is new?") and its intensity ("how big is the change?"). The word "innovation" can refer both to the outcome—something has changed—and to the process—a change is under way.

Innovation Means Change

The "change" part of this definition differentiates innovations from inventions. Inventors think about new ideas. They imagine new useful concepts.

They conceive new designs or new approaches that were not known before. But innovators create new realities. They make sure that the concepts, designs or approaches are used, adopted and applied in specific contexts where they are perceived as new.

Albert Einstein invented "E = mc²", but the Manhattan project innovated with the first atomic bomb, and nuclear energy applications were introduced throughout the world by multiple organizations. Similarly, Leonard Da Vinci invented many things (such as the airplane), but most of his inventions were implemented only centuries after he died.

Innovators Versus Inventors

This means that many inventors fail to become innovators—their ideas are never implemented. It also means that all innovators are not inventors—you can in some cases successfully implement somebody else's new idea.

Numerous patents are granted for inventions which ultimately do not lead to any applications. Many large and very successful businesses actually manage innovation by efficiently putting on the market concepts initially invented by others.

Innovation Versus Invention

This "change" aspect also means that innovating differs greatly from inventing; the former requires different skills and involves different people. Innovations are collective processes, the outcomes of team and organizational efforts and know-how, rather than of individual thinking. Innovations also unfold over time as a continuous process; they do not appear spontaneously ("Eureka!"). Finally, innovations are "subjectively" new, as perceived by specific people in specific situations: what is new to one person might not be new to another. In contrast, inventions are seen as "objectively" new vis-à-vis the state of the art in their field.

The Internet, the personal computer and high-speed trains are innovations resulting from combinations over time of multiple inventions by multiple organizations using new and existing tools and infrastructures. They were not invented overnight by one person, even if some inventors and some inventions played a key role.

So What?

Many organizations fail to innovate not because they do not have enough good ideas, but because they fail to scale them up, get traction and make them happen. Inventors must be convinced, but innovators need to be convincing. Innovation must be seen as 1% inspiration and 99% perspiration.

Kodak famously invented and patented digital camera technology but was ultimately disrupted by the resulting innovations. Airbnb, Uber and Google have successfully implemented concepts actually invented by others, and the initial inventors of the concepts have themselves long been forgotten.

2.2.2 Newness Is Relative

Newness refers to the degree to which something is perceived as different from what existed before. Whether something is new and the degree to which it is new are therefore subjective judgments influenced by relevant persons and contexts.

The electric car is often considered a new concept by customers and managers in the automobile industry, although it has existed for more than a century. New concepts such as mobile telephony or additive manufacturing were introduced decades before they gained visibility. The degree of newness of a "new" model of civilian aircraft is often perceived differently by passengers, pilots, airlines or airports.

Newness Is Relative

Therefore, what is considered "new" by one person can be considered a simple imitation or adaptation by another.

Most of the Fables *by Jean de la Fontaine were actually modified versions of existing stories, some of them centuries old. The "sharing economy" business models launched in the 2010s, such as Airbnb, can be seen as an extension of the C2C online marketplaces launched during the 1990s by companies such as eBay. The homepage of the Google search engine and the "look and feel" of Apple iPhones have only incrementally changed over the last decade, although both firms are very successful and considered among the most innovative companies.*

New for Whom?

A firm might try to apply technologies it masters in a new market (i.e. to the firm the technologies are old while to the market they are new), act as an imitator and adopt technologies that are already known elsewhere (new to the firm, old to the market) or develop innovations that are new both for itself and for its environment (new/new). In particular, this subjectivity of newness offers significant untapped opportunities to international firms, as what is known in one local market might be perceived as new in another.

Most process innovations in the food or packaging industries are invisible to customers (new/old), while payment technologies that have been used for years in some regions are completely new to others (old/new). As a consequence, international consumer goods firms such as ABInBev or Unilever can create significant value by rolling out their existing portfolio of brands and products across multiple geographical areas.

The Liability of Newness

The degree of newness of an innovation as perceived by an organization and its environment is often associated with its riskiness and difficulty (the "liability" of newness, Freeman et al. 1983). The more something is perceived as new by many people, the less we know about how they will actually deal with it. This has practical implications in particular for market research, whose reliability tends to decrease with the degree of perceived newness of an innovation.

How firms, regulators and customers will deal with new/new innovations such as self-driving cars or personalized drugs is still largely unknown. While cash is universal, various payment and "fintech" innovations (mobile payment, contactless, etc.) have met with widely varying degrees of success across regions or countries.

So What?

When dealing with an innovation, a manager should carefully consider whether and how much the innovation is perceived as new by his or her organization and its various constituents, by the environment in which it operates (market, industry, country, etc.), or both, and manage the innovation accordingly.

2.2.3 Many Small Steps and a Few Big Bets

Change can be defined as the "process of becoming different", at individual, organizational or social level. It can relate to modifications of both perceptions—new ways of thinking and feeling—and realities—new events and behaviors. In this context, an innovation can be defined as a change toward something considered new by those experiencing the change.

People often consider "new technology" that which did not exist when they were born and that they therefore had to adjust to. The so-called natives (such as "digital natives") tend to see such changes very differently.

How Big Is the Change?

From an innovation point of view, an important aspect is the intensity of the change as perceived by those experiencing it. This intensity can range from simple improvements (incremental innovations) to full revolutions (radical innovations).

Incremental innovations relate to "small" or marginal changes: improvements, enhancements or modifications of the way things are done while keeping the general frameworks and structures in place. From a business point of view, it is "business as usual": improving competitiveness by introducing small variations in some components or aspects of the firm's management. Thus incremental innovations tend to occur frequently and are often perceived as low-risk initiatives.

Examples of incremental innovations are the frequent cost- and time-saving adjustments implemented in many service organizations or in industries such as in the automotive and chemical sectors. Other examples include the continuous adjustments in the websites of the major e-business players.

Big Jumps

Radical innovations relate to "big" changes: major or structural modifications, new ways of doing things that make previous approaches obsolescent. Radical innovations challenge and reconfigure existing frameworks and structures (Tushman and Anderson 1988). Rather than continually improving existing practices and routines along existing performance indicators, radical innovations relate to breakthroughs and discontinuities. Radical innovations thus tend to happen infrequently and are often perceived as involving higher risks.

Radical innovation examples include the introduction of vaccines and jet planes, the switch from gasoline to electric engines in cars or the introduction of quantum-based computing and so-called blockchain technologies.

From Continuous Improvements to Revolutions

In practice, specific innovations will range from very incremental (continuous improvements) to fully radical (revolutions), depending on the point of view considered. While less visible than radical innovations, the bulk of economic benefits actually comes from incremental innovations and improvements.

Whether the introduction of high-performance triple-glazing or vacuum glazing in construction is an incremental or radical innovation will depend upon whether you are a window manufacturer, an architect, a tenant or a real estate investor.

Incremental innovations can also provide opportunities for equipment and service providers to "jump on the bandwagon" linked to a radical innovation, by providing improved tools for others to try to develop new revolutionary products. While the new products might fail, the tools will still be used.

During a gold rush, it is often more profitable to sell shovels than to look for golden nuggets.

So What?

A key issue for managers is therefore deciding when to pursue a specific technology trajectory(manage incremental innovations) and when to "pull the plug" and switch to a new trajectory (manage radical innovations). The former means risking obsolescence and being overtaken by new trends, while the latter means risking irrelevance and wasting resources by going too far or too quickly.

2.3   Innovation as a Process: Beyond Ideation

Innovation is 1% ideation and 99% perspiration. Most innovations require end-users and various stakeholders to change, and most people do not like to change.

The first job of the innovation manager should therefore be, on the one hand, to identify the key individual and organizational stakeholders linked to the adoption of an innovation and, on the other hand, to understand how much the innovation is potentially disruptive to their individual and collective routines.

Next, the innovator should strive to make it worthwhile and easy to change for all the key internal and external stakeholders. This means trying, on one

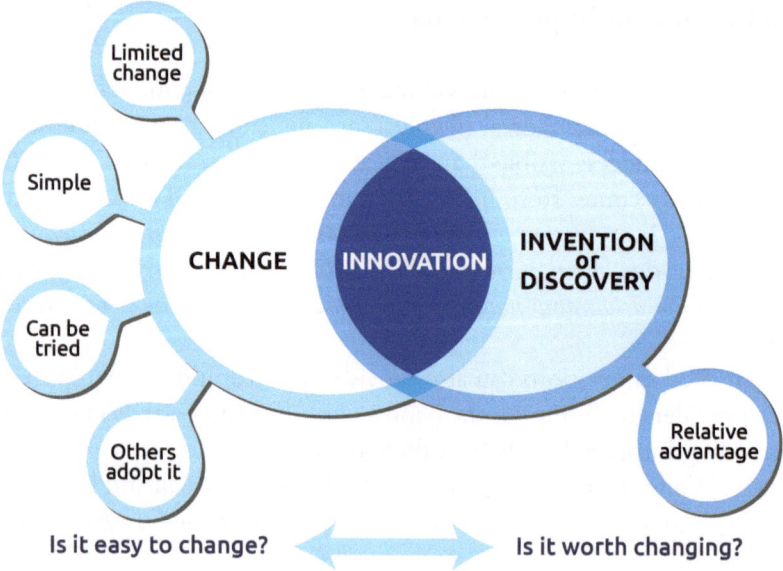

Fig. 2.4 Triggering the innovation process. (Adapted from Rogers 2010)

hand, to maximize for each stakeholder the perceived benefits related to the innovation and, on the other hand, to minimize the perceived costs and risks associated with its adoption (Fig. 2.4).

Inventors are convinced, but innovators must be convincing.

Key Insights

i. Most people *resist change*. As a consequence, the main job of an innovator will be to drive adoption, by convincing people and organizations to *disrupt their routines*.
ii. Driving adoption means demonstrating to key stakeholders that *disrupting the status* quo is worth it and that adopting the innovation will bring significant perceived benefits.
iii. Driving adoption also means convincing key stakeholders that adopting new routines will be neither too difficult nor too risky, and that they and others can *easily make it happen*.

2.3.1 Resistance to Change: Disrupting Routines

One of the main reasons firms fail when managing innovations is that they underestimate the amount of change the innovations imply and consequently the amount of time and energy required to make the resulting changes happen

(Hannan and Freeman 1984). While people often underestimate the long-term impact of innovations and their unexpected applications (they cannot imagine revolutions), they consistently make overly optimistic assumptions regarding how much and how quickly innovations will actually mature and be adopted (they confuse hype and reality).

Since the introduction of the first credit cards more than 60 years ago, people have often predicted the disappearance of cash as a means of payment. But 20 years after Coca-Cola unveiled its first vending machines that accept mobile payments, billions of dollars worth of traveler's checks are still used and thousands of ATMs are still being installed. Even after bitcoins, coins and banknotes are still here, probably to stay.

People Resist Being Changed

People need time and energy to be convinced to adopt innovations because most people fear change. On the one hand, most people see change as unattractive and therefore not worth it. They suffer from loss aversion: they overestimate the cost of losing what they have today, and therefore often prefer to stick to the status quo. On the other hand, they also see change as costly and risky, and therefore too difficult. They are uncertain about their ability to adopt unfamiliar ways of doing things, which might not work or be challenging. Changing means disrupting routines and routines tend to induce strong inertia.

From Socrates's disdain for written texts (and preference for oral tradition) to Barack Obama's criticism of iPods and iPads, each new communication mode (novels, cinema, comics books, videogames, etc.) has initially been vilified. As another classic example, bicycles were once proclaimed hazardous, because they allowed young people to more easily escape their parents' supervision and potentially flirt with strangers.

Organizations Resist Being Changed

The challenge is far greater when whole organizations rather than individuals have to change. Organizations entrench routine, establishing patterns of collective behavior that are often mindlessly followed (Becker 2004). Such routines allow organizations to economize cognitive resources (routines make things simple), reduce uncertainty (provide stability), foster knowledge accumulation (impart experience) and facilitate coordination and control (establish legitimacy and balance of power). Abandoning them is therefore often perceived as risky and costly, and requires significant time and energy.

In 1900, a famous expert predicted that by 2000 people would buy home-delivered manufactured meals from automatic machines, whole cities would be air-conditioned and all transportation would be underground. Barcodes were introduced in the 1970s but are still not used in many industries. In 1989, the OECD predicted that by 2000 optical computers, connected objects and artificial intelligence would be widely used.

So What?

The implication for innovation managers is that they should consider themselves to be change agents much more than creative minds. The implication for innovative organizations is that change itself should become a routine, and that the time and effort required to make change happen should not be underestimated, as is too often the case.

2.3.2 Drivers of Adoption: Make It Worth It

The first way to facilitate the change implied by an innovation and therefore to facilitate its adoption is to make change worthwhile. This means convincing potential adopters that the new way of doing things is more attractive than the status quo owing to the benefits, performance, features or value-added it will provide. While this might be in theory obvious from a marketing point of view, guessing how people will value an innovation often remains challenging in practice.

The first smartphones were aimed at businesspersons for travel or information purposes, but their initial success was actually driven by teenagers who used them to play games, download ringtones and listen to music. The first smartwatches and "smart glasses" were commercial flops because most consumers did not see the point beyond their supposed "cool" value, while professional users embraced them for new industrial applications.

Selling the New Value

The perceived value of an innovation can be promoted through traditional communication channels, dissemination and endorsement. It can also be fostered by competitive intensity, as multiple firms push for an innovation. In some cases the number of adopters is itself a source of perceived value, through so-called network effects. In this case, a virtuous circle of accelerating adoption can be triggered: the more people adopt, the more other people want to adopt.

Apple is a great example of a firm that managed to convince potential customers of the amazing value of all their products, via, among other means, consistent branding and careful orchestration of its product launches. Its AppStore application marketplace is also an example of positive network effects.

Managing Adoption Phases

An important aspect of the perceived value of an innovation is that individuals value "newness" in different ways as an innovation unfolds (Stock et al. 2016). In the initial "introduction" phase, innovations are typically adopted by more adventurous users, who tend to perceive newness as a good thing per se. In later, more "mature" stages, skeptical or conservative users tend to adopt only when they see stronger evidence of benefits.

Crossing the Chasm

While this "adoption phases" approach is product-dependent and sometimes considered too simplistic, it provides valuable insights regarding how the perceived value of an innovation can evolve as it gets adopted, and how adoption by some users might influence others to adopt. In particular, how much the adoption by initial users can be leveraged in order to influence potential later adopters can therefore be a key accelerator to consider and manage. In contrast, an initial success (among more adventurous users) can often be misinterpreted as evidence of future wide adoption (including by more conservative users) and lead innovation into a "chasm" (Moore 2014).

It is one thing to sell expensive electric cars to wealthy customers who want to look "cool and green". As was experienced multiple times by various incumbent and would-be car manufacturers over the last century, it is another thing to reach mass-market adoption.

Dealing with Fears

The perceived value of innovations, in particular those based on new and complex technologies, can be strongly affected by the evolving social consensus regarding key facts (the potential consequences of adoption) and key implications (the prospective benefits of adoption). How much potential adopters will take into account "precautionary principles" (the belief that pre-vention is required even when serious or irreversible damage is not certain) when faced with new societal challenges is often very difficult to predict.

Technology-intensive innovations such as nuclear energy applications, geneti-cally modified organisms or nanotechnologies have undergone wide variations regarding their perceived risks and public acceptance. Similarly, social perceptions regarding car accidents, privacy issues and smoking vary significantly across coun-tries and over time.

So What?

Managers need to convince potential adopters that changing is worth it, tak-ing into account their appetite for newness and fears of uncertainty. They especially need to balance the perceived benefits of an innovation, its potential positive and negative uses as well as the perceived risks it generates. Finally, they need to try to anticipate how users will deal with these risks.

Big data and artificial intelligence offer great opportunities in terms of conve-nience and personalized services. They also raise tough questions and fears regard-ing privacy, democracy and freedom of information. The jury is probably still out.

2.3.3 Drivers of Adoption: Make It Easy

A positive perceived advantage, even a significant one, is in most cases not enough to guarantee the fast adoption of an innovation. What is also needed to facilitate and accelerate adoption is limiting the perceived costs and risks related to the changes it implies.

From cost-effective energy-saving investments to convenient electronic payment methods, there are numerous examples of innovations whose perceived benefits are clear but whose adoption remains slow or non-existent. Meanwhile, innovations with less obvious benefits but which were very easy to adopt, such as text messaging or the Rubik's cube, have been widely successful in a very short time.

Minimizing the Switching Costs

On the one hand, the perceived *costs* related to the adoption of an innovation will be limited if users are convinced that they actually do not have to change much in order to adopt it. An innovation will therefore be quickly adopted if it is compatible with the existing standards, infrastructures, regulations, culture and routines of the people and organizations involved. The perceived "switching cost" will be low if potential adopters are convinced that the innovation is not disrupting their values, norms, practices, mental models and experience.

An innovation will also be easier to adopt if it does not disrupt "the order of things", the meaning and negative or positive connotations associated with new technologies, as well as the "dominant design", (Anderson and Tushman 1990) that is how specific products (like planes, trains, cars or phones) are supposed to look and operate in the mind of users.

A hydrogen-powered or hybrid car is driven and refueled essentially like a gasoline car (it is "compatible"). It also looks a lot like how a car "is supposed to look" (the "dominant design"). But a fully electric car requires a new recharging infrastructure and new ways of planning journeys, and could take a completely different shape. Even emergency services (e.g. dealing with a crash) and pedestrians (who are used to associating silence in a street with safety) will have to adapt. As another example, some hospital patients will be scared by so-called nuclear medicine (which has a negative connotation) but not afraid to eat fully organic and locally grown (positive connotations) but maybe poisonous mushrooms.

Keep It Simple, Stupid

As a consequence, innovators should often understate the "revolutionary" aspects of their innovation in order to foster adoption. They should disrupt competitors but not customers. They must also understand whether their innovation fits in with the existing standards and "ambience" and therefore how much they need to shape it or adapt to them. As Einstein supposedly said: "Everything should be made as simple as possible, but no simpler".

Creating Social Pressures

The perceived risks related to the adoption of an innovation will be limited if users can "see themselves using it". This means that adoption and diffusion (Rogers 2010) will be fostered by "simplicity" (the innovation is easy to understand), "triability" (possibility to experiment) and "observability" (possibility to see whether others are adopting or not).

Insulating a home is often perceived as a complex, irreversible and unpopular investment, although it is in many cases much more efficient than installing solar panels. But people can see when all their neighbors install solar panels, not when they insulate their homes.

Innovation Bottlenecks

The wide adoption of an innovation requires in most cases that multiple stakeholders, and not only the end-users, are convinced of its positive cost/benefit and relatively low risks. And any of these stakeholders could become a bottleneck for the adoption of the innovation. Unlocking adoption bottlenecks therefore requires innovation managers to identify the key internal and external stakeholders, understand the factors most likely to facilitate or hinder their adopting the innovation and, finally, decide how to adapt the innovation and its implementation in order to deal with those factors. For innovation managers, convincing internal stakeholders (e.g. colleagues with sales, purchasing, manufacturing, finance or regulatory responsibilities) as well as external ones (distributors, advisers, partners, etc.) is therefore often as critical as convincing end-users.

Renault initially struggled with its famous "Espace" model because none of its dealers wanted to have a "van" in their showroom. Similarly, new mobile payment services face adoption bottlenecks because they need to align the expectations of financial institutions, telecom operators, hardware and software suppliers, regulators, merchants and users.

So What?

Innovation managers must carefully identify the key internal and external stakeholders linked to their innovations and understand what will make adoption both worth it and not too difficult or risky for each group. They then need to adapt and manage their innovation accordingly, in order to both maximize and accelerate adoption.

2.4   Innovation Typology: Beyond New Products

Innovativeness is often measured in terms of R&D spending and sales from and/or number of launches of new products. But a firm can innovate in many other ways than by launching new products derived from its R&D (Fig. 2.5).

From an operational point of view, firms can—and should—innovate and be more competitive by continuously improving their processes and products. They can and should also improve the value propositions they design around those products and add new services to their offers.

From a strategic point of view, firms can also innovate and be competitive in new ways by introducing new sources of differentiation, implementing

Fig. 2.5 Innovation typology

new business models or building distinctive innovation management capabilities.

Key Insights

i. Innovation is about both making new things *("what")* and making similar things in new ways *("how")*.
ii. Innovation is about *new value propositions and new services*, new ways to market to and interact with customers. This means much more than developing new products.
iii. Innovation is also about finding new ways to differentiate, and new *value curves that disrupt* competitors but not customers.
iv. Innovation is ultimately about designing *new business models* and new ways to deliver, share and capture value.

2.4.1 What Is New: What Versus How

The traditional way of classifying innovations is distinguishing product innovations (putting new goods or services on sale, or for short: new "what?") and process innovations (producing and delivering existing goods or services in new ways: new "how?") (Utterback and Aberbathy 1975).

Doing New Things

Product innovations are what many people actually associate with "innovation". Launching new or improved products can allow firms to enter new markets and gain or protect market share in existing ones.

Engineering and consumer goods companies such as Siemens and Nestlé have been continuously introducing new products for nearly a century, building and developing strong businesses along the way.

Doing Things in New Ways

Process innovations are innovators' best-kept secrets. While in many cases invisible to end-users, they are everywhere. Process innovations allow firms to stay competitive and profitable through efficiency gains, be it in terms of cost, flexibility or waste. While process innovations have historically focused on manufacturing processes, they now include a whole range of technical and organizational innovations applied in various functions, from R&D to purchasing, manufacturing, distribution and advertising.

While the water, gas or electricity delivered to your home is probably quite similar to what your grandparents enjoyed, the way it is generated, treated, distributed and sold has completely changed in recent decades.

Organizing Things in New Ways

Process innovations also include "managerial" or "administrative" innovations, such as new business practices ("just-in-time"), new ways of working and recruiting (co-working, 360° evaluations), managing knowledge (data mining) or even financing new businesses (crowdfunding).

Digital technologies have allowed the introduction of new consumer products such as smartphones and GPS systems. But the true digital revolution is happening within firms, completely changing the ways their products are designed, delivered and sold.

New Things in New Ways

Let us stress that while product and process innovations are distinct concepts, they tend to happen simultaneously, as most product innovations involve some process innovations and as many process innovations directly or indirectly affect the resulting products. Similarly, the product innovation of one firm can become the process innovation of another.

The launch of a new product by a firm (product innovation) will often involve some adaptation of its manufacturing and marketing processes (processes innovations). Similarly, a firm can launch a new 3D printer or a new drone (product innovations) that will be used to improve the operations of another firm (process innovations).

New Products and New Services

The introduction of new or improved offers in the service sector should also be considered a "product" innovation ("doing new things"). Services are known to be by nature intangible, information intensive and based on interactions. New services are therefore often considered to be challenging to develop in a systematic way and to protect from imitators. But firms with the right human resource and information management approaches have shown that new and improved services can be introduced in an effective and systematic way.

Retailing firms such as Amazon, Walmart and McDonald's have demonstrated that basic service industries could be transformed through innovation. Even traditional sectors such as banking or professional services have engaged in "product" innovations in order to remain competitive.

So What?

Innovation managers must make sure that they carefully balance their investments in time and resources between doing new things (new products or new services) and doing the same things in new ways (new processes or new organizations). While there are many other ways to innovate, continuously innovating in these "core" dimensions remains a key issue in many industries.

2.4.2 New Value Propositions and New Services

Marketers and salespersons have long understood that customers do not just "buy a product". What they do is reach a purchasing decision at a certain moment, in a certain setting, under a given format, with a certain level of personalization, for a given price, in order to fulfill a perceived need, complete a job or solve a problem. All these dimensions offer opportunities for firms to innovate through new value propositions or through new services, even if the good actually delivered can often remain essentially the same.

New Marketing

Innovative value propositions can include new, customized or improved packaging, design or interface (e.g. single doses, recyclable or reusable packaging), new pricing models (such as leasing, freemium or free advertising-based offers) and new types of product promotion and placement (such as new product bundles or search- or location-based advertising).

Mineral water companies are in many cases forbidden by regulations to change their product in any way, but it did not prevent some of the most famous brands from developing very innovative value propositions. Similarly, a new windshield ordered a few months in advance and delivered in bulk to a car manufacturer does not create the same value as an identical windshield delivered in a new way, in less than 24 hours, in customized packaging, to a local repair shop.

New Markets

"Marketing" innovations also include pitching existing products to new customer segments, or repositioning an existing offer in line with new or emerging needs. Leveraging features or perceived value in such new ways allows firms to reach new customers with what is essentially the same product or service, hence reducing development and industrialization risks and costs.

From newborns to expats to divorced couples, retail banks have found multiple new customer segments to target with simple variations of their basic offers. Similarly, some pharmaceutical companies have discovered that what used to be considered the side effect of a drug could actually become a key buying factor (e.g. an anti-depressant drug that also makes patients stop smoking).

New Interactions

The second avenue for developing new value propositions that go beyond new or improved goods is to realize that the interactions between a firm and its customers throughout the product lifecycle are also sources of value-added that can be leveraged in innovative ways.

Firms provide value-added to customers through what the customers purchase but also through multiple interactions before, during and after purchases. While these interactions are often seen as necessary solely to support the sales of goods, they actually also offer opportunities to develop innovative services related to the sold goods. These new services can be offered to potential customers, customers or users for free, as differentiating factors, or become themselves a source of revenue for the firm.

Open-source software providers actually offer their products for free but generate revenue through specialized advice, installation, maintenance, training or customer support services bundled around the free product before, during and after the actual purchase of the software. As another example, Nespresso and Starbucks have shaken the coffee industry through innovative bundles of services around what remains essentially a cup of coffee.

Servitization

Pushing this approach further, the sales of a product can actually be completely replaced by the sales as services of the functionalities initially offered by the product ("product-as-a-service"). In this case, ownership is replaced by use and customers purchase competence, features and capabilities rather than physical goods. As Theodore Levitt said decades ago: "People don't want to buy quarter-inch drills; they want to buy quarter inch holes".

Innovative product-as-a-service examples include industrial lighting (rented), aircraft engines (leased), computer storage and software (cloud services and SaaS), music (streaming), razors and mobility (shaving subscriptions, car and bike-sharing).

Leveraging Servitization

Such "servitization" approaches have two important implications for innovation managers. On the one hand, they generate significant human resource and information technology challenges, especially for firms used to "simply" delivering goods.

On the other hand, the customer intimacy generated by the resulting continuous interactions with users can be a very valuable source of market intelligence. As a consequence, while competitive differentiation tends to decrease with product sales (the more a firm sells, the more competitors will try to copy it), it actually increases with service sales (the more a firm sells, the more it knows about its customers and can serve them better).

Online retailers like Amazon have to meet huge challenges in terms of data management, customer services and logistics. But they now know so much about who buys what and when that they can design their value propositions better than their competitors.

So What?

For most innovation managers, developing marketing and service innovations should be considered at least as important as developing technology innovations focused on products and processes. Moreover, mastering the resulting interactions and leveraging the "big data" they generate is becoming a sustainable source of competitive advantages in many industries.

2.4.3 New and Disruptive Value Curves

Developing new products, value propositions and services and developing new processes to manufacture and deliver them allow firms to remain competitive. But innovations can also allow firms to develop new ways to compete, by leveraging new and disruptive sources of differentiation. While more risky, new types of value propositions can allow some firms to innovate from a strategic point of view, escaping the "red oceans" of low-margin competition and capturing significant value.

Low-cost airlines did not introduce new aircraft, new frequent flyer programs, new meals, new seats or new pilots to compete with traditional airlines. They introduced new "value curves": new value propositions that incumbents find much more difficult to replicate.

New Value Curves

Traditional approaches to competition use benchmarks to help firms identify which features of their product and which performance indicators (speed, weight, shelf life, etc.) they need to focus on. In contrast, "value innovation" approaches differentiate according to other features (existing or new), beyond the focus of existing competitors. They aim to introduce new value propositions that might be less effective in some dimensions but could provide better value in others that matter to enough customers.

Harley Davidson was struggling to compete on price, power and reliability with Japanese motorcycle manufacturers. It survived by competing in different ways and focusing on other features such as customization and engine sound. As a counter example, Nokia, Kodak and Blackberry, which could not directly compete with the iOS and Android software platforms, tried to differentiate themselves by introducing new smartphones with features such as far greater camera resolution or enhanced privacy protection, but they struggled to get sufficient customer traction.

Disruptive Innovations

Pushing this logic further, firms can introduce innovative value curves linked to so-called disruptive innovations (Christensen and Bower 1996). These target features, such as convenience, low cost and simplicity, that matter to low-end or new customers, rather than the high-end functionalities and performance required by more sophisticated customers. They aim to be "good enough" for most customers rather "the best", which often results in overshooting customers' expectations.

Music or video streaming, MP3 format and digital cameras were initially regarded by the music, television and film industries as low-quality, low-margin alternatives. Professional users and sophisticated amateurs indeed initially rejected them. But the general public loved them. In contrast, many high-end cars and consumer electronic products have "gadget" features that allow them to outperform competitors in benchmarks but have limited value to most customers.

Disrupting Incumbents

Disruptive innovations are a key threat to incumbents because even well-managed firms can be threatened without being outperformed. Well-managed incumbents that use traditional marketing approaches will tend to focus on high-margin and sophisticated customers and develop high-performance products. But because of this high-end focus, they might not pay enough attention to alternative, unproven and/or initially low-margin approaches. As a consequence, they will often react to the threats posed by disruptive innovations only when it is too late.

Leading watchmakers initially considered Swatch a low-quality product owing to its being unrepairable and its limited reliability. But once 400 million "swatches" were sold, including to high-end customers, and once several of these watchmakers had been acquired by the Swatch Group, it was too late for these traditional incumbents to react. In another example, firms producing all-in-one Satnav devices such as TomTom disrupted the in-car systems developed by automobileO-EMs. Then the all-in-one devices were in turn disrupted by initially less sophisticated but more convenient smartphone apps.

So What?

The emergence of innovations based on new and disruptive value curves means that managers cannot afford to focus only on competitive benchmarks and operational excellence. Focusing only on existing competitors and customers and current key performance indicators can lead to missed opportunities to innovate. It also means being at risk of disruption by someone who leverages such opportunities.

2.4.4 New Business Models

In recent decades, Internet and digital technologies have revolutionized the way firms exchange goods and information, both internally and with their partners and customers. This has created opportunities to define new "ways of doing business", new ways of generating and capturing value. While some of these "business model" innovations (Amit and Zott 2001) predate the emergence of the Internet, the ability to deliver online services to a worldwide audience with nearly zero marginal costs has fostered their multiplication.

The Key Elements of a Business Model

A business model can be defined as the specific answer of a firm to two questions: What are the value proposition and the target customers (product/market positioning)? How is the value proposition delivered to the customers in a profitable way (value chain)?

If a firm wants to develop a business that focuses on eradicating malaria, it could serve customers such as the inhabitants of poor or rich countries, the actors in those countries' healthcare systems, local and international non-governmental organizations or public institutions. It could focus on vaccines, treatments, mosquito nets and repellents, or offer new ways to control mosquitos through better water management or manipulating mosquito DNA. While all these options would potentially create value by eradicating malaria, they involve completely different product/market positioning and value chains.

Innovative Business Models

Most firms focus on improving individual elements of their existing business model, developing new value chains on the one hand and new value propositions on the other hand. In contrast, firms developing business model innovations introduce whole new configurations of both their market positioning and value chain, new "stories" regarding how to answer the two questions stated above (Chesbrough and Rosenbloom 2002). Such approaches can be considered the simultaneous combination of different types of innovations, introducing new ways to create and capture value.

Rather than improving how they design, manufacture and sell vehicles, some car companies have introduced as a business model innovation an online platform that allows users to seamlessly rent and use different types of private and public transport. Similarly, a traditional retailer could offer to potential customers the ability to buy via auction goods or services from a wide range of amateur and professional suppliers. Finally, a payment company could provide its services for free, making money by selling to third parties the transaction data it collects. None of these cases involve radically new processes or products. What is radically new is how they are combined.

New Ways to Innovate

Ultimately, firms can develop new business models by being innovative in the way they manage innovation itself. They can develop new ecosystems and strategic approaches or build new innovation capabilities to identify, select and capture opportunities.

New ways to innovate include rapid-throughput screening and biomimicry approaches in R&D, crowdsourcing and open innovation approaches in new product development, corporate venturing and shared infrastructures, new risk management and learning approaches as well as product lifecycle and agile manufacturing approaches.

So What?

This means that innovation managers should consider not only the development of new resources and technologies, but also the development of new configurations of them, new "ways of doing business". Similarly, when developing a new technology, managers should focus not only on improving the performance of that technology but also on the design of distinctive business models around it. One valuable source of inspiration for such innovation can be the business models already developed in adjacent industries.

A company developing a new artificial intelligence algorithm could consider selling its software as a product or as a service, selling advice and recommendations defined using the algorithm, using the algorithm to design better offers in various industries or to make profitable trades or investments. The same technology can lead to many potential business models.

2.5 Innovation Strategies: Beyond New Product Development

Strategy is about what we want to be as an organization, today and tomorrow, and how we want to do it better than others. Those questions can be addressed at corporate level, for the firm as a whole, and at business level, in each area where the firm wants to compete (Fig. 2.6).

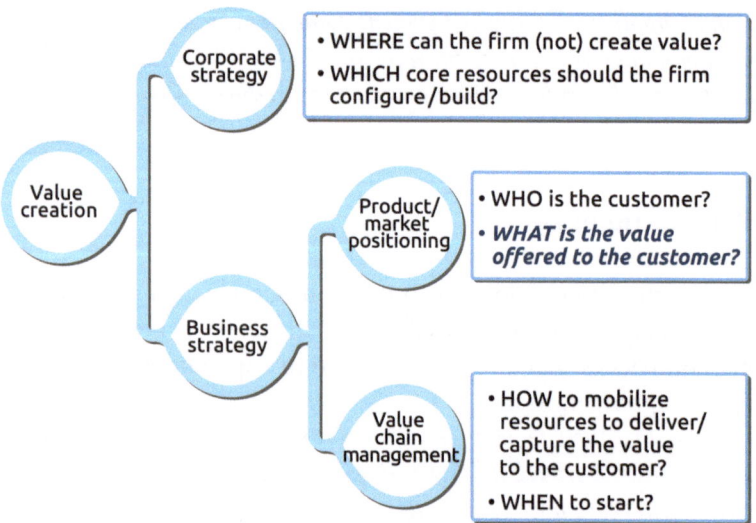

Fig. 2.6 Strategic innovation levers: beyond new product development

The firm's corporate strategy relates therefore to where it wants to compete and create value—in which business areas—and which core resources it will configure or build in order to compete and create value more effectively than others do. The firm's business strategy relates to how it wants each of its businesses to compete, which customers to target, what value to offer and how to more effectively deliver and capture value.

Innovation strategies can be defined as new answers to these questions. In particular, developing more new products and developing them more quickly is definitely not the only effective innovation strategy.

Key Insights

i. The first way to develop an innovation strategy is to (re)define where the firm wants to compete, what it defines as being part of its (new) *"core business"*. Specifically, this means finding new ways to explore—not just exploit—and reconfigure its business portfolio.

ii. The second way to develop an innovation strategy is to redefine *how the firm wants its businesses to compete* and which product/market positioning and value chain it wants them to build and sustain.

iii. A key element of an innovation strategy is to understand that *first is not always best*. A firm should assess and define how quickly it wants to enter new business areas and manage innovation accordingly.

iv. One high-risk/high-potential innovation strategy is to challenge generic strategies and define and implement a *"strategic innovation"*, a radical redefinition of where and how a firm wants to compete and shape its future.

2.5.1 Innovative Corporate Strategies: (Re)define the Core Business

The first strategic question regarding innovation is for the firm to define or redefine what it sees as its "core business" today and tomorrow. A firm's "core business" is an ambiguous concept but can be defined as what its key stakeholders see as its "legitimate territory" as a firm. It refers to the "market spaces" where the firm believes it can leverage a unique set of resources, capabilities, know-how, routines and processes, giving it unique strengths over its competitors. It should also explicitly refer to what the firm chooses "not to do".

Farming or Hunting

Given the global and specific trends affecting a firm's activities, a key trade-off in terms of innovative corporate strategies is therefore to decide when to "farm" and exploit the current business portfolio—play by the rules—and when to "hunt" and explore new potential core businesses—set new rules (Greve 2007). Too much "farming" and the firm will miss new potential synergies. Too much "hunting" and the firm will exhaust its corporate resources.

Pepsi and Coke stopped many years ago seeing their strategic priority as only "selling more cola than the other". They each redefined their business portfolio, including other types of drinks (juices, mineral water, etc.) and even other types of products (fast-food). As a counterexample, Blockbuster once refused to link its "core business" of brick and mortar video-rental activities with the new online business of Netflix, and eventually faced bankruptcy. Finally, Lego nearly went bankrupt through excessive corporate exploration, and then refocused its core business to the theme of its slogan "Obviously Lego but never seen before".

Winning the Game

"Farming" or "exploitation" corporate strategies can be related to incremental innovations. They focus on refining existing business models in order to maintain and improve their performance. Their objective is to "win the game" by developing sustainable competencies, defending existing businesses and markets and picking low-hanging fruit. The keywords of such corporate strategies are execution, compliance and adaptation. A key element of effective innovative corporate strategies based on exploitation is to maintain a balanced allocation of resources across the various businesses of the corporate portfolio, taking into account their respective growth potential and attractiveness as well as their current market share and competitive strength.

Disney has developed over decades a core business built around unique capabilities to manage creative talent and studios, and to "farm" its unique content across television channels, music studios, merchandizing, publications and comics, movie theaters and amusement parks.

Changing the Game

"Hunting" or "exploration" corporate strategies can be related to radical innovations and creative destruction. They focus on entering new business areas and finding new ways to perform. Their objective is to "change the game" by leveraging cross-industry projects with partners and being ready to make bold moves and cannibalize existing businesses. The key words of such corporate strategies are experimentation, leadership and shaping the environment.

Innovative corporate strategies based on exploration can focus both on "extending" and "reshaping" the core business. Extending the core means looking for adjacent opportunities to grow the business portfolio and finding new ways to leverage corporate assets. Reshaping the core means designing new value-added business portfolio configurations, for example, through specialization, consolidation, integration or internationalization. Extending can be seen as diversifying existing business portfolios, reshaping as creating an entirely new business portfolio "shape".

Telecommunication operators all over the world have discovered that geographic expansion, data transmission (the Internet) and now entertainment (television, streaming, etc.) are business areas that they could (and should) enter. These areas, which were initially "new business", have in most cases become "core business", replacing their historical activities (voice communications) and in some cases reshaping their entire business portfolio.

So What?

At corporate level the first innovation strategy questions to answer should be: Who are we and what do we do as a corporation? What are the business areas we want to occupy today and tomorrow? Then managers should identify the core capabilities they need to build or acquire as a consequence, and how to efficiently balance those scarce capabilities between the exploitation of the current core business and the exploration of potential new core businesses.

2.5.2 Innovative Business Strategies: Redefine How to Compete

Once a firm's "core business" and the resulting business portfolio have been defined or redefined, the next strategic question regarding innovation is how to compete in each of the firm's businesses. This means deciding whether the firm wants to develop a new product/market positioning and/or a new value chain.

According to Michael Porter's famous model, generic business strategies include "niche" and "cost leadership" strategies, which imply different ways to manage innovations. Niche strategies aim mainly to develop economies of scope and enhanced quality and features, such as precision or quality, targeted at customer niches. Cost leadership strategies aim mainly to compete on costs through minimum and standardized features, economies of scale and efficient logistics and manufacturing. In the first, product/market innovations (what do you sell to whom?) will be key, while in the second case process innovations in the value chain (how do you deliver value?) will often matter more.

What Do You Sell to Whom?

The first business strategy question is therefore, "What does the business sell to whom?" (product/market positioning). In terms of innovation, the firm must decide whether its business should stick to its current product ranges and potentially streamline them. It could also develop variations (new products, new services, new ways to differentiate, etc.) or try to design completely new offers. Conversely, the firm must decide whether the business should focus on its current markets, target adjacent markets or try to replicate its success in completely new market segments.

Most banks offer a combination of payment and intermediation, insurance and advice services to customer segments such as general consumers, affluent consumers, small and medium enterprises, corporations and institutions. Some banks have, however, chosen to develop innovative product/market positioning around specific subsets of services and/or customer segments (e.g. not offering mortgages or selling only third-party investment products) or around new ones (e.g. new services to expatriates).

How Do You Deliver Value?

The second business strategy question is, "How will the business deliver and capture value?" (value chain). In terms of innovation, the firm must decide whether its business should stick to its current position in its industry value chain or move to a new one. A business could move and/or integrate upstream (e.g. toward B2B approaches) or conversely downstream (B2C). It could try to bypass existing players (disintermediation) or create new intermediary positions (such as digital platforms).

Once the position of a business in the industry value chain has been defined, the firm must also decide whether to stick to its current "business" value chain and focus on improving its efficiency (e.g. through digitalization) or develop its activities in new ways. This means deciding how the business will manage its key activities, such as the design of its value proposition (R&D, marketing, etc.), operations (procurement, logistics, etc.) and customer management (sales, customer services, etc.), while providing the necessary support and governance (management, legal, human resources, etc.). This includes deciding not only how to perform these activities but also whether to "make or buy" them, through targeted partnerships or in the open markets.

All airlines will fly passengers from A to B. But each will choose how much it integrates and interacts with other players (e.g. aircraft manufacturers, airports or travel agents). Each airline will also choose whether and how the activities required to deliver and capture value will be performed in-house, with partners or on the open markets. A passenger might buy a ticket from one firm, have his or her baggage handled by another, receive catering from a third and fly in a plane operated by a fourth. Some airlines have even developed completely new businesses, such as catering or ticketing services, out of specific industry value chain positions.

So What?

Finding new ways to compete means analyzing in a systematic way what the business could offer, which needs it could fulfill and which value-added and support activities it could focus on. Each business should consider what should be its (new) product/market positioning and its (new) industry and "business" value chain. This includes new ways to design and deliver its value proposition but also new ways to organize its business and interact with its customers. Developing new products is indeed only one among many possible innovative business strategies.

The pizza industry has seen retailers, convenience stores, fast-food actors and online intermediaries develop new business strategies while still delivering essentially the same combinations of mozzarella, tomatoes and pepperoni.

2.5.3 Timing: First Is Not Always Best

When an opportunity to develop a new business strategy has been identified, a key strategic decision concerns the timing of the implementation of that strategy. Can and should the firm be the first mover? If there already is a first mover, is it worth being a "fast follower"? Is "wait and see" a valid strategy? Decades of academic research have demonstrated that there are no obvious answers to these questions (Lieberman and Montgomery 1988). But what is certain is that first is not always best. There are advantages and disadvantages to both being and not being first.

The "first movers" of most of the software people use every day (word processors, spreadsheets, web browsers) are firms that are long dead and forgotten. The first advertising-based search engine was not developed by Google; the first fast-food hamburger was not sold by McDonald's, the first disposable nappies were not Pampers and the first MP3 player was not the iPod.

Run Ahead

There are obviously some advantages that can be captured by the firm that is the pioneer in an industry. It can build brand loyalty and technological leadership, be ahead in the learning curve and secure intellectual property. It can also pre-empt scarce assets such as critical natural resources, locations or exclusive partners. The pioneer can also leverage switching costs and increasing returns advantages from the customer base it has built before the others.

Once most customers associate a product with a specific firm (such as Google, Velux or Bic) or get used to a specific location or design (the Amazon home page or the Qwerty keyboard), dislodging them from their entrenched position can be very difficult.

Dominant Design

If there is already a strong consensus in the market regarding how a product should look and work (a "dominant design", Anderson and Tushman 1990), it is often more profitable to try to improve the way this product is delivered, through process innovation, rather than trying to dislodge the existing offer through product innovations. In other words, if there is already a "first mover" in terms of product, a firm can still try to be first mover in terms of process. In some cases the subsequent improvements in an innovation after its first introduction may be vastly more important, economically, than the innovation in its original form.

Samsung has initially built its success on manufacturing cheaper and better versions of products others had already developed, such as microwave ovens or flat-screen televisions. Rocket Internet has built its strategy around the effective replication in new geographical areas of existing (mostly US-based) Internet business models.

The Second Mouse Gets the Cheese

There are also some obvious disadvantages linked with being the first mover. The biggest one is free-riding by fast followers, who learn from what worked and did not work for the pioneer. Fast followers can benefit from the spillovers of past R&D expenses and infrastructure development, poaching the key employees of the first movers or inventing around their intellectual property. Fast followers can benefit from all the efforts the first movers spent on educating buyers, understanding their requirements and coping with uncertain technological choices.

Apple had to explain to consumers what a "tablet" was, to the subsequent benefit of Samsung. Tesla had to convince drivers that a green car could be cool, and build a recharging infrastructure that other car manufacturers will be able to leverage. The initial genome project had to find out the general structure of human DNA, to the benefit of its followers.

When the First Will Be Last

First movers in a specific technology will also often be less well-positioned when a new technology replaces it, because of its path dependency. Past first movers often have the most to lose, through cannibalization. They also often have the biggest investments locked in technology-specific assets such as supply and distribution networks and industrial assets. As a consequence, fast followers are often more profitable and sustainable than first movers.

Some pharmaceutical companies still stick to outdated production processes because "it works" from a regulatory and compliance point of view, and because they have invested huge sums in their legacy manufacturing capabilities. In other words: "Pioneers get the arrows, settlers get the land".

Pacing Innovation

Managers considering whether to develop a new business strategy should "manage the hype" and understand that there is no such thing as an absolute first-mover advantage. They should assess the pros and cons of the available options (first mover, fast follower or "wait and see") based on the characteristics (complexity, barriers to entry, network effects) of the business they are considering. They should then maximize the benefits and manage the drawbacks of their position, for example, by investing in secrecy, patent protection, complementary assets or customer lock-in.

So What?

Innovation managers need to be smart movers, not necessarily first movers. Moving too quickly means having to cope with costly uncertainties and resistance to change. Moving too slowly means risking obsolescence owing to new trends and disruption. And not deciding or deciding too slowly whether or not to be the first is probably the worst option. Innovation is about pace, not speed.

2.5.4 Off the Beaten Paths: Strategic Innovations

When managers face mature markets where available adjustments to their business strategy have already been implemented and fast-follower benefits have already been captured, they might see no obvious ways to avoid price wars and commoditization. When technological improvements and generic strategic opportunities such as cost leadership or differentiation have already been captured, finding new ways to develop a business strategy can seem a dead-end.

Diving into "Blue Oceans"

One high-risk but high-potential business strategy option is to develop a "strategic" or "value innovation" approach (Chan and Mauborgne 2004). Such an approach combines a first-mover advantage and an "exploration" posture with a complete redefinition of the business. This does not mean adjusting the product, the market or the value chain but rather simultaneously reconfiguring all three. It means no longer trying to extract value from existing competitors, suppliers and customers but rather making such competition irrelevant by offering fundamentally new ways to create, capture and deliver value.

The car industry used to be stuck with incremental innovations and price competitions among incumbents, all offering roughly the same range of small/cheap, mid-range and large/powerful/expensive cars. Then came completely different approaches such as "monospaces" and "SUVs" (Renault Espace, BMW X5), small but expensive or large but cheap cars (Smart, Mini, Dacia), and green but cool cars (Prius, Tesla), which could not fit in the existing range and changed the competitive game. Similarly in the financial sector, "Fintech" start-ups try to disrupt the businesses of incumbent financial institutions through radically different ways to deliver and capture value.

Strategic Innovations

In a "strategic innovation" approach, the industry environment is no longer seen as a given set of constraints to adapt to but rather as something which can be changed and shaped. Competition is no longer about increasing existing sources of value and/or decreasing the cost of the value chain but rather about completely redefining it. Strategic innovation also means trying to capture first-mover advantages by occupying new competitive spaces, what W. Chan Kim called new "blue oceans".

So What?

Managers considering such a radical, first-mover and "exploitation" approach must of course weigh their potential benefits in terms of new sources of profitability against all their drawbacks in terms of uncertainty, delays and ambiguity. In particular, they need to ensure that they have the resources to explore and capture unproven and often fuzzy business strategies, which might remain unprofitable for several years while requiring the costly development of new capabilities.

Pharmaceutical firms, having seen their pipeline of blockbuster drugs dry up and facing price pressures from regulators, must find new spaces in which to operate and new ways to create value. But this means restructuring their core assets and capabilities and engaging in risky and costly ventures.

2.6   Drivers of Innovation Strategies: Beyond Hype

The first and one of the most important questions a firm should address before developing its innovation strategy is "Why do we want to innovate?" Innovation is not a good thing per se; it should not be the first priority for all firms, all the time. Always being the most innovative, in all dimensions, is never a guarantee of long-term corporate success.

The way a firm decides how to manage innovation must therefore be aligned with the unique resources it can leverage, acquire and develop, the

Fig. 2.7 Defining why you want to innovate

competitive and market environment it faces at regional and global level, and the expectations and purpose of its key stakeholders (Fig. 2.7). What was right for one (even very successful) firm in terms of innovation might not be the best option for another.

Key Insights

i. Innovation is *a means, not an end* per se. A firm's innovation strategy should be about how its management of innovations will help it to leverage its unique resources and adjust to changes in its environment, in line with its corporate purpose.

ii. Innovation strategies should be based on what the firm understands are the present and future *key success factors* relevant in the environment it faces, including in particular new partnerships, and its present and future *sources of competitive advantages*, including in particular its unique datasets.

iii. Innovation strategies should be based on a shared understanding of who are the *key stakeholders*, including shareholders, employees, customers and society, and what are their *expectations*, especially regarding the time horizon considered and level of ambition.

iv. Strategy should drive innovation, not the other way around. A firm's *innovation posture* should therefore be defined based on what its strategic objectives are in terms of innovation, and not just on the latest innovation buzzwords or recipes.

2.6.1 Innovation as a Means, Not an End

Being the most innovative, all the time and in all dimensions, is never a good strategy. Successful firms are not the most creative ones, the ones spending the most in R&D or the ones launching the greatest number of new products. Successful firms are those that can manage innovations to leverage their strengths, cope with their weaknesses, seize opportunities and escape threats (the famous "SWOT" framework), all in line with their values and purpose. Managing innovations well always matters; always being the most innovative never does.

A quick check of any ranking based on business performance (profitability, growth or valuation) will reveal many successful firms such as Exxon, GE or J&J which are not, or are not known as, always being the most innovative ones. And some firms once perceived as innovation leaders, such as Enron or Vivendi, have not always been corporate successes. Conversely, firms such as Coca-Cola or Ferrero have been very successful over decades while not being particularly notable in terms of innovation. As Theodore Levitt already noted in 1966, innovation is not a "great tribal god" that all managers should worship.

What the Firm Needs to Do Well

The first objective of a strategy based on innovation should be escaping obsolescence and/or irrelevance, capturing opportunities and dealing with threats. Innovation management must help firms identify new competitive positioning in their environment and deal with the latter's evolution, including global and specific trends such as new industry or regional structures and networks, changes in competition and profit pools, and market and technological developments.

Nokia did not fail because its technology or organization lost effectiveness. It failed because the key success factors in its industry changed in a way to which it did not manage quickly enough to adjust.

What the Firm Can or Could Do Better Than Others

The second objective of a strategy based on innovation should be escaping commoditization and price competition through differentiation, and enabling the firm to leverage its strengths and cope with its weaknesses. Innovation management must help firms to carefully but effectively leverage their resources in order to build their competitive positioning better than others. These resources can include technical and organizational capabilities, current cash flows and funding sources, reputation, customer base and values as well as underused assets.

Mid-size automobile firms such as BMW cannot predict or control the evolution of their industry and markets, but they can continue to leverage and innovate around their key assets in terms of technologies (existing and new), design and brand.

What People Want from the Firm

The third and last (but not least) objective of a strategy based on innovation should be maintaining the firm's legitimacy, or its "license to operate", based on its purpose and the expectations and values of its key stakeholders. Innovation management must help firms on the one hand achieve the right level of performance while dealing with risks and uncertainties, and on the other hand remain aligned with their corporate values and vision.

Firms dealing with products that affect climate change, ecosystems or consumer health (such as oil or tobacco companies or industries based on genetically modified organisms) see their success as driven as much by social acceptance and legitimacy as by their competitiveness.

So What?

The first question an innovation manager must be able to answer is "Why does the firm want to innovate?" Innovation is not an end per se. It is a means for firms to develop new strategies aligned with their unique assets, the new key success factors in their environment and the expectations and risk-aversion of their stakeholders.

2.6.2 Key Success Factors and Sources of Competitive Advantages

The management of innovation will be aligned with the strategy of a firm if it can help to identify and implement new and better ways to deal on one hand with the (new and existing) key success factors in its environment and on the other hand with its (new and existing) sources of competitive advantages.

(New) Key Success Factors

First, the (new) key success factors of a firm will be driven by the global and specific policy, economic, socio-cultural and technology trends affecting the competitive ecosystems in which the firm operates, and how competitors, customers, suppliers, new entrants, partners and substitutes deal with these trends. New regulations, new technologies or new customer needs will affect the relative competitive strengths of each player. In particular, the ability to effectively cooperate (and not only compete) with other players is becoming a key success factor of value creation.

The competitive ecosystems of leading Internet players such as Google, Amazon, Facebook and Alibaba are constantly evolving. The customers of yesterday become the competitors of today and the suppliers or partners of tomorrow. New technologies such as artificial intelligence and connected objects can change the balances of power and bring in new competitors and/or partners such as IBM or Siemens.

(New) Sources of Competitive Advantage

Second, the sources of competitive advantages of a firm will be driven by the necessary and unique resources it can acquire and develop (Barney 1991). This includes tangible resources (assets, cash, natural resources, data, etc.), which are often relatively easy to replicate, more intangible resources (technology, reputation, partnerships, etc.) and "organizational" resources (skills, social capital, energy and drive, etc.), which are often less easy to replicate.

These resources and their combinations can allow the firm to develop unique competencies, privileged assets and special relationships and to protect those capabilities against imitation, substitution, hold-up or slack. Strategic resources that can be leveraged for innovation include business platforms, untapped and/or unique customer insights and underused technological or marketing capabilities.

The huge sets of data some firms can now generate and manage have become sources of competitive advantage that they can leverage in their industry or in new ones. While this creates significant risks in terms of security (hacking) and privacy, it also generates significant new opportunities in terms of process efficiency, decision-making and knowledge management. As John Naisbitt said, "Running out of this resource [data] is not the problem, drowning in it is".

New Kids on the Block

New sources of competitive advantages also allow new sources of competition to emerge. Free-riders can use new business models to turn the business of a firm into a commodity, by leveraging other sources of profit (such as advertising or data). Disruptors and strategic innovators can bypass industry incumbents by developing less effective but more convenient or completely different value propositions, based on new capabilities.

Steel, automotive and finance are examples of industries that have seen the emergence of strong new entrants that build competitive advantages based on new types of resources, and that come from completely different horizons than their "traditional" competitors.

Learn to Learn

Finally, faced with a changing environment and new competitors, a key source of competitive advantages for firms is to develop what David Teece (2007) and others call "dynamic capabilities", or the ability to "learn to learn", and to continuously improve their abilities and rapidly adapt to changing internal and external conditions.

The challenge in quickly evolving industries is no longer to "plan and execute right". It is now to be less wrong than the others and to be able to quickly learn and adjust to new advances.

So What?

Corporate innovations do not happen in a vacuum. Innovation managers must understand not only what a "good" innovation could be, but also which innovations are "the right ones" for a given firm. This means understanding how innovations could help the firm cope with new key success factors and how much they are leveraging its sources of competitive advantages. It also means detecting the new competitive threats these innovations create and building the organizations capacity to continuously learn about them.

2.6.3 Purpose: Managing Stakeholder Expectations

Like innovation, staying competitive is a means, not an end. The management of innovations will therefore be aligned with the strategy of a firm only if it can meet the expectations of its key stakeholders, especially regarding its level of risk, ambition and time horizon.

Alphabet's (formerly Google) "moonshot" projects or the space travel start-ups of some leading entrepreneurs can afford to invest billions in cash as long as the main stakeholders of the firms involved are ready to pursue their ambitious but long-term and very uncertain objectives. But history shows that such "tolerant" expectations can be unsustainable.

Identifying the Key Stakeholders

The key stakeholders an innovation strategy should consider include the customers and users (in terms of value, reputation and satisfaction), the shareholders (in terms of valuation, profitability, growth or cash-flow impact), the managers and employees (in terms of engagement and employer brand) and current and future generations (in term of societal and environmental impact). While many claim that the expectations of stakeholders can ultimately be aligned in the long term, they clearly represent trade-offs when considering innovation management strategies.

Whether outsourcing, automating and/or digitizing existing value chain activities is a valuable innovation strategy will depend on whether you are a shareholder, employee, customer or neighbor. Those innovation strategies might be valuable for all those stakeholders in the long run, but as John Maynard Keynes famously said: "In the long run we are all dead."

Setting the Clock

The second element of an innovation strategy that needs to be clarified regarding purpose and expectations is the time horizon. Given the time and energy it takes to turn a new idea into a new reality, the way innovations can be managed is directly affected by the time horizon. In particular, a firm must decide whether and how much it wants to focus on defending its existing businesses (short term), growing or acquiring emerging businesses (mid-term) or exploring new technologic platforms (long term).

The effects of climate change will be global and sizeable only in decades or even centuries. But consumers and businesses are reluctant to make innovative energy-saving investments with payback times longer than a few years. Similarly, a strong corporate investment in radical innovation initiatives will typically pay back in only five to ten years, while immediately affecting quarterly cash flows and outlasting most CEO tenures. Finally, a new drug, a new energy source or a new aircraft design can take decades to become profitable.

Sizing Expectations

Once the relevant time horizons have been defined, the third element of an innovation strategy that needs to be clarified regarding its purpose and expectations is comprised of the level of ambition and the corresponding risks the stakeholders are ready to take. Key stakeholders must be aligned regarding the potential value creation they are pursuing and the implications in terms of risks and investments. There is too often a wide gap between the performance improvements expected from innovations and the time, risks and resources the firm is ready to deal with.

One way to try to align expectations is to reverse-engineer what the gap is in terms of growth or profitability between ongoing and existing projects (such as expected market growth or planned M&A projects) and what is implied in the share price and/or in the growth target of the firm. The corresponding risks and investment required can then be derived using past corporate projects with similar ambitions as proxies.

The valuation of some Internet-related firms implies (sometimes quite heroic) assumptions regarding their ability to develop new innovative businesses.

This can help align stakeholders regarding what they really want (sometimes including conflicting or ambiguous objectives) and manage their expectations regarding what it will imply (often much greater investments, delays and risks than what they are ready to cope with). As Paul Saffo said, "Most ideas take twenty years to become an overnight success".

Even asset-light Internet businesses such as Google or Airbnb took five to ten years to go from a project to a sizeable (but not necessarily profitable) business. For innovative but very large corporations such as Procter Gamble or 3M, even growing yearly sales by a few percentage points can be a challenging objective, involving significant risks and investments.

So What?

Before launching an innovation management strategy, managers should clarify who the targeted stakeholders are, their time horizon and their level of ambition. They should also align expectations regarding the level of investment and risks these imply. In other words, if you do not know where you want to go, you are unlikely to get there.

2.6.4 Define Your Innovation Posture

While innovation is increasing in importance for a growing number of firms, the right way to manage innovations is not trying to be as innovative as possible all the time, whatever that means. The right way to manage innovations is to define and implement an innovation strategy, that is, to choose where, how and how much to innovate, in line with the firm's resources, environment and objectives. Sometimes it might even be a good idea not to innovate, or not to innovate too much and too quickly.

What is remarkable about corporate champions such as Google and Apple is not the frequency or the intensity of their innovation efforts, but rather how carefully they choose when and when not to innovate and how carefully they implement these decisions. They have been successful because they have executed very, very well a limited number of innovation initiatives, not because they launched a great number of such initiatives.

Pick Your Innovation Mix

As a consequence, firms should choose the right mix of innovations (radical versus incremental; what versus how; first mover versus follower) and the right type(s) of innovation (process, product, offer, business model, etc.) to pursue in support of their strategy. They then need to identify the capabilities they need to maintain or develop in order to identify, assess and capture these innovation opportunities.

A firm focusing mainly on boosting the profitability or sales of its existing businesses ("adapt and evolve" or "shape") should in most cases focus on a large number of incremental process innovations ("how"), implemented across its businesses and improving short-term performance and operational effectiveness. It should also aggressively pursue "fast-follower" approaches based on industry and market intelligence.

A firm focusing mainly on building new businesses or new options for future businesses ("build" or "renew") should in most cases focus on a small number of radical product innovations ("what"), implemented in dedicated structures, improving long-term value and leveraging first-mover advantages.

These two extreme cases illustrate that there are very different ways to effectively manage innovations, and that picking the right one is essential.

Maybe "Uberization", crowdsourcing, lean start-up or open innovation approaches are what a given firm needs, given its resources, environment and objectives; but maybe not. "There is no 'perfect' strategic decision," Peter Drucker has said. "One always has to pay a price. One always has to balance conflicting objectives, conflicting opinions, and conflicting priorities. The best strategic decision is only an approximation—and a risk".

So What?

Innovation managers should identify the right innovation "posture" for their firm and the right innovation strategy given its current and future environment, resources and objectives. It might be the best option for some firms to follow defensive, imitative or traditional approaches, while others should be more offensive or opportunistic. The key success factors of innovation postures will not be their innovativeness but their innovation management capabilities and their alignment and consistency across the organization.

2.7   Synthesis

Build a Shared Strategic Vision of Innovation: Key Insights

2.1. Why it matters: innovation management capabilities

i. Megatrends such as technology disruptions, international competition and sustainability affect firms across all sectors and industries, creating new competitive and social challenges.

ii. Industries and sectors are also disrupted by new regulations, new customer needs and new technologies, forcing firms to reconsider the sustainability of their assets and activities.

iii. Small and large firms across sectors must place innovation among their strategic priorities if they do not want to suffer the fate of the dinosaurs.

iv. This implies developing innovation management capabilities to identify, select and capture the right innovation opportunities, in line with the firm's ecosystem and strategy.

2.2. Innovation as a business: more than creativity

i. Innovation means much more than invention. Managing innovation means managing both newness and change, and the latter often matters the most.

ii. Newness is relative. What is today new to one manager, its organization or its environment might not be to another.

iii. Innovation is about changing people's perceptions and realities, combining many small steps and a few big bets.

2.3. Innovation as a process: beyond ideation

i. Most people resist change. As a consequence, the main job of an innovator will be to drive adoption, to convince people and organizations to disrupt their routines.

ii. Driving adoption means demonstrating to the key stakeholders involved that disrupting the status quo is worth it and that adopting the innovation will bring significant perceived benefits.

iii. Driving adoption also means convincing the key stakeholders involved that adopting new routines will be neither too difficult nor too risky, that they and others can easily make it happen.

2.4. Innovation typology: beyond new products

i. Innovation is about both making new things ("what") and making similar things in new ways ("how").

ii. Innovation is about new value propositions and new services, new ways to market to and interact with customers. This means much more than developing new products.
iii. Innovation is also about finding new ways to differentiate, and new value curves that disrupt competitors but not customers.
iv. Innovation is ultimately about designing new business models and new ways to deliver, share and capture value.

2.5. Innovation strategies: beyond new product development

i.   The first way to develop an innovation strategy is to (re)define where the firm wants to compete, what it defines as being part of its (new) "core business". Specifically, this means finding new ways to explore—not just exploit—and reconfigure its business portfolio.
ii. The second way to develop an innovation strategy is to redefine how the firm wants its businesses to compete and which product/market positioning and value chain it wants them to build and sustain.
iii. A key element of an innovation strategy is to understand that first is not always best. A firm should assess and define how quickly it wants to enter new business areas and manage innovation accordingly.
iv. One high-risk/high-potential innovation strategy is to challenge generic strategies and define and implement a "strategic innovation", a radical redefinition of where and how a firm wants to compete and shape its future.

2.6. Drivers of innovation strategies: beyond hype

i.   Innovation is a means, not an end per se. A firm's innovation strategy should be about how its management of innovations will help it to leverage its unique resources and adjust to changes in its environment, in line with its corporate purpose.
ii. Innovation strategies should be based on what the firm understands are the present and future key success factors relevant in the environment it faces, including in particular new partnerships, and its present and future sources of competitive advantages, including in particular its unique datasets.
iii. Innovation strategies should be based on a shared understanding of who are the key stakeholders, including shareholders, employees, customers and society, and what are their expectations, especially regarding the time horizon considered and level of ambition.
iv. Strategy should drive innovation, not the other way around. A firm's innovation posture should therefore be defined based on what its strategic objectives are in terms of innovation, and not just on the latest innovation buzzwords or recipes.

Bibliography[1]

Amit, R., & Zott, C. (2001). Value creation in e-business. *Strategic Management Journal, 22*, 493–520.

Anderson, P., & Tushman, M. (1990). Technological discontinuities and dominant design. *Administrative Science Quarterly, 35*, 604–634.

Barney, J. (1991). Firm resources and sustained competitive advantage. *Journal of Management, 17*, 99–120.

Becker, M. C. (2004). Organizational routines: A review of the literature. *Industrial and Corporate Change, 13*(4), 643–678.

Chan Kim, W., & Mauborgne, R. (2004). Value innovation: The strategic logic of high growth. *Harvard Business Review, 82*(7–8), 172–180.

Chesbrough, H., & Rosenbloom, R. S. (2002). The role of the business model in capturing value from innovation: Evidence from Xerox Corporation's technology spin-off companies. *Industrial and Corporate Change, 11*(3), 529–555.

Christensen, C. M., & Bower, J. L. (1996). Customer power, strategic investment, and the failure of leading firms. *Strategic Management Journal, 17*(3), 197–218.

Freeman, J., Carroll, G. R., & Hannan, M. T. (1983). The liability of newness: Age dependence in organizational death rates. *American Sociological Review, 48*, 692–710.

Greve, H. (2007). Exploration and exploitation in product innovation. *Industrial and Corporate Change, 16*(5), 945–975.

Hannan, M. T., & Freeman, J. (1984). Structural inertia and organizational change. *American Sociological Review, 49*, 149–164.

Lieberman, M. B., & Montgomery, D. B. (1988). First-mover advantages. *Strategic Management Journal, 9*. Special issue: Strategy content research (Summer, 1988), 41–58.

Moore, G. A. (2014). *Crossing the Chasm, 3rd Edition: Marketing and Selling Disruptive Products to Mainstream Customers.* New York: Harper Collins.

Rogers, E. M. (2010). *Diffusion of innovations.* New York: Simon and Schuster.

Stock, R., von Hippel, E., & Gillert, N. (2016). Impacts of personality traits on consumer innovation success. *Research Policy, 45*, 757–769.

Teece, D. J. (2007). Explicating dynamic capabilities: The nature and microfoundations of (sustainable) enterprise performance. *Strategic Management Journal, 28*(13), 1319–1350.

Tushman, M. L., & Anderson, P. (1988). Technological discontinuities and organizational environments. *Administrative Science Quarterly, 31*, 439–465.

Utterback, J. M., & Aberbathy, W. (1975). A dynamic model of process and product innovation. *Omega, 33*, 639–656.

---

[1] An extended bibliography is available at www.NavigatingInnovation.org

Bibliography

3

Manage Entrepreneurial Ecosystems

A perfect innovation strategy is worthless if the firm does not have the organizational and entrepreneurial abilities to execute it. Innovations can succeed only when people and teams across organizations, networks and whole ecosystems embrace change and make new things happen. This means being able and willing to experiment, learn and often fail. As managers this also means being able to let others experiment, learn and sometimes fail.

The second innovation management challenge is therefore to foster the right entrepreneurial behaviors at all levels: people, teams, organizations, networks and regional ecosystems (Fig. 3.1).

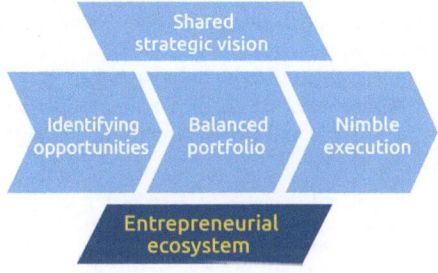

3.1 Encourage people to innovate: corporate entrepreneurs
3.2 Build and lead effective innovation teams: balancing acts
3.3 Build and manage innovation-ready organizations
3.4 Develop innovative networks and collaborations
3.5 Create innovation ecosystems
3.6 Synthesis

Fig. 3.1 Managing entrepreneurial ecosystems

© The Author(s) 2018
B. Gailly, *Navigating Innovation*, https://doi.org/10.1007/978-3-319-77191-5_3

3.1 Encourage People to Innovate: Corporate Entrepreneurs

When trying to capture innovation opportunities, the main barrier most firms encounter is not lack of ideas or lack of funding but lack of the right people: people who have the motivation and ability to make things change and to face the inherent risk and uncertainty. People who are ready and willing to act like entrepreneurs, and managers who are ready and willing to let them.

Those "corporate entrepreneurs" are made, not born. With the right triggers, anybody can choose to engage in identifying, assessing and capturing new opportunities, as well as mobilizing resources and gaining legitimacy (Fig. 3.2). The challenge for managers is therefore to foster the right entrepreneurial behaviors, by using levers like corporate culture and values, rewards and training as well as attitudes and norms, in order to create the right perceptions within their organization.

Key Insights

i. Innovation is made by people. But most innovators are not natural born entrepreneurs. They are not lone heroes creating single-handedly revolutionary products. They are ordinary people, often within existing firms, who decide to change things and are ready to cope with the resulting uncertainties.

ii. Corporate entrepreneurs do not plan, do, check and act like traditional managers. They focus on finding and mobilizing available resources in order to identify, assess and launch new initiatives, while adjusting goals and expectations along the way.

iii. Entrepreneurial intentions and behaviors can be managed and fostered. Tolerance for failure, reward systems, role models and norms, as well as skill development and slack resources can create the right attitudes, norms and perceptions.

3.1.1 Entrepreneurs: Made Not Born

Innovation can happen only when some people decide to make new useful things happen, to create "new combinations" (Joseph Schumpeter), to act "out of the box" and to "leap forward in the face of uncertainty" (Henry Mintzberg). They can be stand-alone entrepreneurs, project managers formally appointed by a firm or "corporate entrepreneurs" (Burgelman 1983), employees launching bottom-up informal initiatives.

Fig. 3.2 Corporate entrepreneurs: ordinary people doing innovative things

Innovation requires these people to challenge the status quo, both in terms of resistance to change—social inertia—and competitive pressures—economic forces.

The first Internet entrepreneurs needed both to convince customers, managers and investors to disrupt their routines and to dislodge or bypass powerful "brick-and-mortar" incumbents. A manager developing a new product in a firm will have to convince both his/her colleagues to adapt their systems and procedures and his/her own salespeople to risk cannibalizing their existing sales.

Made, Not Born

The biggest myth about such "entrepreneurs" is that they are special people born with unique skills and talent, lone superheroes launching revolutions. Some people are indeed more likely than others to act like entrepreneurs, if they have a strong need for achievement, high risk-taking propensity and internal locus of control (they think that what happens to them is mainly driven by their own decisions, not by circumstances or fate). But decades of attempts to select, through psychological testing, future entrepreneurs have dismally failed. And a lone entrepreneur left on a desert island would not achieve much on his or her own. As the entrepreneurship scholar Bill Gartner wrote, "Who is an entrepreneur?" is the wrong question.

Entrepreneurship is a social behavior, not a personal characteristic. In the right circumstances anybody is susceptible to acting like an entrepreneur, driven by necessity or by opportunity.

Someone can be a bureaucrat at the age of 20 and a successful entrepreneur at 40. Another person can be a boring administrator in the office and an exciting social entrepreneur at home. Many successful start-up entrepreneurs used to work in large firms, the same large firms which often complain that they lack entrepreneurial talent. As the scholar Donald Kurtako wrote, "Entrepreneurs are ordinary people who do extraordinary things".

So What?

Corporate entrepreneurs can play an important role in innovation, but they are made, not born. The implication for managers is that the priority should not be selecting or "poaching" future entrepreneurs. To foster corporate entrepreneurship, the priority should be routinely identifying who among current employees and managers are the most likely to engage in entrepreneurial behaviors, and triggering the conditions for them to actually do it.

3.1.2 Corporate Entrepreneurs Versus Managers

Innovation can be managed only if managers and employees do more than brainstorm and think "out of the box". Innovation initiatives can emerge if managers and employees can continuously identify and generate new innovation opportunities, but also assess and develop them, launch and implement the best ones and finally mobilize the right resources to gain legitimacy.

Innovation Politics

Innovations disrupt the order of things and create winners as well as losers. Innovation managers need therefore to build networks and create new coalitions within and across their organization, in order to deal with existing routines and balances of power. They need to understand which parties will cooperate or compete with their initiatives, which ones will regulate, conflict with or accommodate their projects.

As the management guru Gary Hamel wrote, innovation managers must be aware that "Management innovation often redistributes power (so don't expect everyone to be enthusiastic)". As a consequence, they should take into account the observation of innovation scholar Keith Pavitt, that "Major innovation decisions are largely a political process, often involving professional groups advocating self-interested outcomes under conditions of uncertainty (i.e. ignorance), rather than balanced and careful estimates of costs, benefits and measurable risks".

Corporate Entrepreneurship

Innovation initiatives require managers and employees to engage in entrepreneurial behaviors, going beyond their job description and organizational silos in order to learn, pursue new opportunities and cope with uncertainties.

Innovators bend rules and organize new coalitions, to try to create new competitive advantages. Traditional managers or "administrators" follow best practices and rules to maintain the firm's competitive advantages. An organization full of administrators would be frozen, but an organization full of innovators would be hell.

Innovators Versus Administrators

"Administrator" types of managers can be found mainly in industries where pre-existing knowledge is a key source of competitive advantage. They focus on allocating resources, based on best practices, benchmarks, predictions and expected returns.

"Innovator" types of managers can be found mainly in industries with a high level of uncertainty, where adaptability is a key source of competitive advantage. They focus on mobilizing and adjusting resources, based on learning and affordable losses.

Administrators follow maps and directions, innovators blaze new trails.

In particular, innovators often have to cope with limited resources and "mak[e] do with what is at hand" (Claude Lévi-Strauss). Rather than simply allocating their budgeted resources, they often need to engage in "entrepreneurial bricolage" (Baker and Reed 2005) recycling unused or untapped resources, self-teaching new skills and creating alternative institutional arrangements.

Uber drivers in emerging countries have to learn how to cope with local infrastructure and vehicles, hostile incumbents, ambiguous and changing regulations.

So What?

The implication for managers is that stimulating innovation activities and behaviors must go far beyond organizing creativity sessions. It implies creating circumstances where people can explore and learn outside their "business as usual" jobs and beyond their organizational unit, in order to identify and generate new innovation opportunities, assess and develop them, launch and implement the best ones and in parallel enroll the right resources to gain legitimacy within their firm.

3.1.3 Drivers of Entrepreneurial Intentions and Behaviors

Innovation can happen when managers and employees choose to engage in entrepreneurial behaviors, disrupting the order of things in order to discover, evaluate and exploit new opportunities. But innovation is risky and complex, and failure can sometimes be very costly. It is therefore important to understand what can trigger such entrepreneurial behaviors and what can increase the likelihood that managers and employees will decide to engage in them.

Entrepreneurial Intentions and Behaviors

Decisions and behaviors are driven by intentions and circumstances. While circumstances are often difficult to control, intentions are influenced by attitudes, norms and perceptions. Understanding these influences can allow managers to assess and manage them, in order to foster entrepreneurial behaviors within their organization.

What's in It for Me?

The first factor to consider is the potential personal benefits or drawbacks from the perspective of individual managers or employees. This concerns their attitude toward adopting an entrepreneurial behavior. How favorable or unfavorable is their appraisal of that behavior? What are the expected potential consequences for them to adopt that behavior? In particular, what are the potential negative consequences for the individuals if they fail?

Innovation managers must cope with the fact, highlighted by the Nobel Prize winner Richard Thaler, that "In many companies, creating a large gain will lead to modest rewards, while creating an equal-sized loss will get you fired". They must therefore follow Thomas Edison's, view: "I have not failed. I just found 10,000 ways that do not work."

What Will Others Think?

The second factor to consider is the perceived impact on the social position of individual managers or employees. It concerns subjective norms and perceived social pressures, that is, the way managers and employees perceive others' opinions regarding their adopting an entrepreneurial behavior. It relates to the question, "How will people who are important to me – colleagues, relatives or friends – react?"

In particular, it concerns the way individual innovators are recognized and rewarded within the organization. What are the metrics used to assess the success or failure of an innovation initiative? What types of feedback, rewards and/or recognition is provided as a result?

Designing the right metrics and rewards is very context-specific and can lead to perverse effects, in particular by promoting individual rather than team performance, and creativity rather than efficiency. But a universal proven approach is to identify and promote innovative role models within the organizations, in particular among the top managers who are ready to "walk the talk".

Another fact is that while extrinsic motivations (rewards and punishments) can sometimes be used to encourage people to change, intrinsic motivations (self-desire) have repeatedly been shown to be more effective at fostering creativity and imagination.

3M, for example, awards "Genesis Grants", worth as much as $100,000, to innovative company scientists for research. Solvay used to celebrate "Innovation Champions" every three years, by identifying and rewarding winning teams in multiple categories.

Is It Feasible?

The third and final factor to consider is the perceived feasibility. This is the perception of individual managers and employees regarding their potential ability to successfully adopt an entrepreneurial behavior. Do they perceive that they have the required skills to perform all the required activities? Do they perceive that they have the time as well as the potential access to the required resources?

In particular, perceived feasibility concerns the provision of training and professional development opportunities (can I learn to act like an innovator?) as well as the availability and access to "slack" resources (Nohria and Gulati 1996) within the organization (can I find the required budget, expertise and time?).

Google is famous for its policy of allowing its employees to spend 20% of their time on pet projects. While this is certainly a great tool in terms of fostering employee engagement, there is limited publicly available evidence of a significant direct bottom-line impact. And when Alphabet (Google's parent company) had to consider strategic innovation projects like its famous "moon shot" projects or Google+, it managed those projects through dedicated structures and acquisitions, not through organization-wide, employee-driven innovation approaches.

So What?

The implication for managers is that the entrepreneurial behaviors needed for innovation within their organization can be fostered. This requires understanding the current hurdles in terms of attitudes, norms and perceptions and applying the right balance of extrinsic and intrinsic motivation, resources and pressure for people to engage in innovation activities.

3.2   Build and Lead Effective Innovation Teams: Balancing Acts

Innovation is made by teams, not by lone individuals. While some people can be very creative and/or be inspiring leaders, it is only when they manage to assemble and motivate teams to act that innovation can actually happen (Fig. 3.3).

Moreover, innovation projects often involve complex and interdependent tasks, to be completed within tight time and resource constraints. As a conse-

Fig. 3.3 Building and leading innovation teams

quence, a "lone wolf" corporate entrepreneur who is not able to become an effective team leader or member will quickly become a distraction or a bottleneck rather than an innovation asset.

Finally, developing winning innovation teams requires a careful balance of pressure and autonomy, focus and diversity as well as the right group dynamics, decision-making processes and social identification.

Key Insights

i. Organizations can successfully manage innovation only if they have the ability to build and lead effective project teams around their innovation initiatives.

ii. An effective project team will become a great innovation team if it can manage key trade-offs in terms of level of pressure, diversity, slack and tolerance for diverging and outside perspectives.

iii. Building an effective innovation team means mobilizing "transformational" leaders and team members with the right profiles, background, motivation and skills—not just whoever happens to be "on the bench" when the initiative is launched.

3.2.1 Why We Need Effective Project Teams to Innovate

Innovation is not only about creative individuals having inspiring ideas. It is about teams and organizations that are able to identify, assess, launch and legitimate new opportunities while facing ambiguous circumstances and commitments. "Normal" projects are already difficult enough to manage in a competitive environment. But innovation-related projects must be able to rely on dedicated and effective teams, given the complexity, interdependencies and pressures involved.

Complexity

Innovation projects involve by definition some level of newness for the organization, its environment or both. They therefore involve technological uncertainties for which past experience and existing processes and routines can be of limited value. Innovation is about uncharted territory, which often leads to complex projects.

The first ever sequencing of human DNA entailed completing a formidable task in terms of volume. But it also struggled because in some cases the team did not even know what it was looking for. As another example, Samsung once had to recall millions of products because the complexity of inserting a new battery in its tightly designed smartphones was underestimated.

(Global) Interdependencies

On top of their intrinsic complexity, innovation projects are often difficult to split into autonomous parts. They tend to mobilize tasks and people with strong interdependencies and from multiple fields (marketing, manufacturing, etc.). This is even more challenging when the tasks and functions are spread across multiple geographies, as is increasingly the case in global firms. Cultural "distance" (Kogut and Singh 1988), local events or competition for resources can turn simple misunderstandings into trust-destroying conflicts. International innovation teams should not rely only on remote technologies to communicate and collaborate.

The development of the Airbus A380 was delayed by several months among others because of incompatibilities between the developments made by its different national teams. In 1999, a space probe destined for Mars crashed because its different international teams had confused metric units (newtons) and "imperial" unit (pounds of force).

Underinvestment

Because firms tend to underestimate the level of change implied by an innovation and because the level of uncertainty involved often hinders the commitment of significant resources ("big bets"), innovation projects typically have to cope with severe constraints in terms of time and resources, that are not aligned with their often ambiguous and ambitious goals. As long ago as the 1950s, researchers such as Edith Penrose identified underinvestment in terms of managerial capabilities and attention as a key barrier to the development of new businesses (Kor and Mahoney 2004).

Ambitious new projects in the aeronautics or energy sectors are famous for being almost always many times over budget and many years late. Many "asset-light" Internet businesses, as opposed to being "overnight" successes, took several years to move from an idea on paper to a sizeable source of revenue.

Dedicated Teams

The complexity, interdependencies and operational constraints most innovation projects have to cope with implies that highly effective project management is a key success factor of innovation. More specifically, it requires mobilizing effective project leaders and sponsors, aligning expectations in terms of roles, deadlines and deliverables, and finally putting in place adequate communication and conflict resolution mechanisms within the team and with its external partners.

Too often innovation projects fail because the team in charge was incomplete or poorly staffed, owing to choosing team members on the basis of availability rather than skills and motivations.

So What?

Building a strong and dedicated team to support an innovation initiative should be a core prerequisite, not a side issue which is dealt with after the decision to launch the project has already been made. In particular, top managers have a key role to play as sponsors, ensuring that the best talents are mobilized throughout the organization. Too often firms discover only after the fact that innovation projects failed because they picked the wrong team to manage them, or because they waited too long before paying sufficient management attention to the project.

3.2.2 How to Build Great Innovation Teams: Key Trade-Offs

What makes a great innovation team can be very context and project-specific, as what works well for one person and one innovation in a particular context might not be adequate for another. Key trade-offs should always be carefully monitored and managed: between being creative and effective, between generating newness and achieving change (West 2002).

Tasks and Objectives

The first trade-off to carefully manage is the balance between assignments (defining the tasks) and commitments (defining the objectives). Innovation teams should have sufficient perceived degrees of freedom and autonomy of means to foster accountability, collaboration and engagement. But they should also have tasks and objectives that are defined clearly enough to be perceived as coherent and achievable.

Fostering innovation is not an excuse for avoiding so-called smart objectives: S, specific; M, measureable; A, achievable; R, realistic; and T, time-based.

Diversity of Knowledge and Skills

The second trade-off to carefully manage is generating sufficient positive friction and cross-fertilization while maintaining a cohesive team. Diversity in terms of background, skills, experience, age, networks and ways of thinking fosters imagination, creativity and problem-solving. It can also help the team cope with the complexity of all the tasks involved in identifying, assessing and implementing innovation opportunities. But too much diversity can also result in a lack of mutual consideration, understanding and trust, and thus generate conflict and haggling.

The diversity involved in the decision-making process of the European Union has often been a source of creative problem-solving. But it has in some cases completely paralyzed its ability to reach a consensus.

Group Dynamics

The third trade-off is to maintain sufficient collective discipline and consistency to "do things right" while allowing sufficient individual initiatives and divergent thinking to "do the right things". The team should be put under enough pressure and supervision to ensure that it will integrate the right operational constraints and converge on its own toward consensus in due time. But the team should also remain sufficiently open to diverging perspectives and tolerate active minorities and dissent. This also includes the ability to liaise with the outside world and integrate external ("not invented here") inputs.

The US-based investment banks are famous for the ability of their individual employees to take risks and effectively compete in a fast-changing environment. Japanese manufacturers are more famous for the efficiency of their teams to continuously improve and capture synergies. The ideal innovation team is probably the right mix of both.

So What?

Innovation teams need not only to be highly effective but also to achieve the right trade-off between their ability to cope with newness and their ability to deliver change. Managers should therefore pay as much attention to how their innovation teams are staffed and managed as to whether they have achieved the expected technology, market or financial deliverables.

3.2.3 Transformational Team Leaders and Members

Great innovation teams must combine great team leaders, who can be formally appointed or informally emerging, and great team members, who can be formally part of the team or de facto regularly collaborating with it. Leaders must be able to go beyond "command and control" approaches, and members must be able go beyond their formal roles and job descriptions, as both are often ill-suited to the uncertainty and ambiguity of innovation projects.

Innovation managers should therefore follow Joan Magretta's advice, based on her experience at Harvard: "The real insight about managing people is that, ultimately, you don't".

Transformational Innovation Leaders

Managers too often see their role explicitly or implicitly as, on the one hand, fixing objectives, allocating and monitoring human and financial resources, and on the other hand as taking corrective actions if these predefined objectives are not met. But this "transactional" style of leadership is not effective when resources and objectives are constantly evolving and when engaging entrepreneurial talents is key. What is needed for innovation is a "transformational" style of leadership (Bass 1990).

Such leadership replaces "command and control" approaches with a combination of inspirational motivation and vision, intellectual stimulation and learning, influence through charismatic role modeling and mentorship, and finally an interpersonal style that promotes autonomy and careful attention to individualized consideration. These are leaders who can simultaneously "walk the talk"—act according to shared values—and "run the race"—deliver performance. Too often the short-term focus is on the latter at the long-term expense of the former.

As the leadership scholar Mary Parker Follett noted already in 1924, "Leadership is not defined by the exercise of power, but by the capacity to increase the sense of power among those who are led. The more essential work of the leader is to create more leaders".

Building the Right Team

The best innovation leaders or corporate entrepreneurs can achieve nothing if they do not manage to build effective teams, by identifying and engaging the right people. They need to combine the right personalities, abilities, expertise and backgrounds, with enough creative thinking and interpersonal skills, in particular sociability. They also need to maintain among team members sufficient intrinsic and extrinsic motivations. Finally, they need to acknowledge that their ability to "manage" great talents is actually quite limited.

Venture capitalists are known to prefer investing in a great team with an average idea rather than in an average team with a great idea. Conversely, conflict within a team, even when reaping great success ("sharing the spoils"), is one of the most frequent hidden reasons for the failure of innovative ventures.

So What?

The key ingredient of successful innovation projects is not technology or money. It is people. The weakest link in an organization that wants to successfully manage innovation is too often the human resources department. Managers must therefore ensure that existing people processes and tools are sufficiently geared toward identifying and developing potential innovation leaders and team members within their organization.

3.3   Build and Manage Innovation-Ready Organizations: How Some Elephants Can Dance

Defining an innovation strategy is useless if the firm does not have an organization able to implement this strategy. It is therefore critical to understand how innovation-ready the organization should be and to efficiently manage the key cultural and structural drivers of organizational innovativeness.

Fig. 3.4 Building innovation-ready organizations

Building innovation-ready organizations goes far beyond building the consumer brands and fancy products that allow some firms to claim the number one spot in many of the innovativeness rankings one can find online. What matters is to put in place the combination of innovation mindset and ventures needed to support the firm's innovation strategy (Fig. 3.4).

Key Insights

i. Innovation-ready organizations are not innovative for the sake of it. They combine operational efficiency with both the corporate culture required to continuously exploit and do things in new ways—innovative mindset—and the corporate structures required to explore and sometimes do completely new things—innovative ventures.

ii. Innovative organizations manage to develop a corporate culture that fosters trust, learning and exchanges in order to do things in new ways and overcome the traditional organizational barriers to corporate entrepreneurship.

iii. Innovative organizations can create dedicated structures where completely new corporate ventures can be nurtured and parented, leveraging their corporate assets while freeing these "teenage" ventures from organizational inertia.

iv. "Ambidextrous" organizations are designed and managed to retain the innovation agility of many small firms, such as flexibility, engagement and autonomy, while capturing the managerial efficiency of large corporations with their scale, assets and power.

3.3.1 Innovation-Ready Organizations: Mindset and Ventures

Innovation is now on the agenda of most organizations. But how to recognize an innovation-ready organization? Does it show the speed and aggressiveness of the All Blacks rugby team or of a Viking ship, the agility and creativity of the Cirque du Soleil or of geeks working in a garage? There are many ways to answer this question. Too often, however, choosing a measure of an organization's innovativeness (e.g. R&D spending, number of patents or notoriety) is driven more by the availability of data than by the measure's relevance to effective innovation management.

Innovation rankings consistently recognize consumer product firms with innovative brand reputations, such as Apple or Google, although their respective organizations are managed in completely different ways and although lesser known B2B firms can be considered more innovative in many ways. Being innovation-ready means much more than building an innovative brand.

Measuring Innovativeness

There are multiple ways to measure the innovativeness of a firm, in terms of inputs, processes and outputs. Innovation inputs include R&D and innovation spending, share of staff involved or number of ideas, projects or partnerships launched. They also include performance gaps identified and shared within the organization. On the other hand, innovation process indicators include cost, failure rate, time to break/even or time-to-market and level of adoption of specific innovation approaches such as patenting or open innovation.

The most innovative sectors in terms of R&D are the ICT, automotive and pharmaceutical industries. Their major players collectively spend more on R&D than most countries do.

Measuring Innovation Outputs

Innovation outputs that can be measured include reputation, number of product launches, revenues from new product sales or licenses, operational improvements and profit or sales growth. They can relate to the novelty and meaningfulness of new products introduced or new market approaches. They can also relate to the introduction within the organization of new production methods, management approaches and technology. Finally, they can relate to innovative behaviors and strategies, such as the willingness and ability to change at individual, team, managerial and corporate levels.

Innovative firms such as Pixar and DreamWorks have built their reputations based on the quality and profitability of the products they decide to launch, rather than on their sheer numbers.

Innovative, but Dead

But being the most innovative, whatever the yardstick, is not the point. Innovation benchmarks can raise interesting questions (why am I faring differently than my peers?) but are not performance indicators, and raising interesting questions and measuring success are quite different things. Innovation-ready organizations combine the corporate culture and structures required to implement both exploitation and exploration strategies. They have the capacity of an orchestra and the mindset of a jazz band.

General Motors was one of the top corporate R&D spenders the year before it nearly went bankrupt. Nokia was for many years ranked among the most innovative firms.

So What?

Innovation is a means, not an end. What matters most isn't using the same level of inputs or the same processes as one's peers. Rather, what matters is to have the level and type of innovativeness that fits with the firm's resources, environment and objectives. What matters is having the innovation-readiness that will support the firm's innovation strategy. This includes the ability not only to do things better (continuous improvements) but also to do things in new ways (innovative mindset) and to do completely new things (innovative ventures).

3.3.2 Doing Things in New Ways: Innovation Mindset

Some groups or populations are known to be consistently better at collectively identifying, assessing and capturing new opportunities. They support creativity and experimentation, risk-taking and bold actions, opportunity-seeking and forward-looking perspectives, individual and team autonomy as well as the right level of competitive aggressiveness.

The UK-based birds such as robins and tits have for decades enjoyed the dietary benefits of the bottles of milk left open in front of country houses. But when dairy distributors started placing aluminum seals on the bottles, only the blue tits managed to learn collectively through experimentation and socialization new ways to access this very rich source of nutrients. Similarly, people who live on large land masses with seamless communications tend to collectively innovate more than isolated islanders do.

Innovation-Ready Corporate Cultures

What innovation-ready groups have in common is the right collective culture. They have acquired over time the right shared basic assumptions, "rules of conduct", vocabulary, methods, rituals, myths and so on, regarding "how things get done here," "how we are expected to behave in specific circumstances." In particular, they share the right organizational culture regarding knowledge-sharing, trust, risk-taking, learning and socializing.

Innovation champions are known to develop organizational cultures where failure is tolerated, a shared vision of innovation exists and people can be patient with ideas. These elements are more important than whether innovation is directly measured ("what gets measured gets done") or whether they invest more in R&D than their peers do.

Fostering Corporate Entrepreneurship

Innovation-ready organizations also create circumstances which foster individual and team corporate entrepreneurship (Amabile et al. 1996), by encouraging individuals and teams to "move around", be proactive and take ownership, be open, learn and focus on new opportunities, tolerate ambiguity, challenge routine, be autonomous, and share and mobilize resources.

Many middle managers perceive a lack of clear direction from top management as paralyzing. Corporate entrepreneurs see such ambiguity as an opportunity to gain new degrees of freedom and shape future developments.

Overcoming Organizational Barriers

Many organizations tend to develop strong "corporate immune systems" against innovation. These include risk-averse decision-making processes as well as unsupportive metrics and rewards that lead to half-hearted funding and bureaucratic project management. They also include governance "silos" that hinder organizational learning and focus on short-term efficiency. Finally, they include "innovation-killer" managers who embody complacency, destructive criticism and fear of failure.

A rigid budgeting process that divides resources among entities through multiple meetings and committees and focuses on protecting quarterly earnings might be effective today but not very innovative tomorrow. When innovation is supposed to be "everybody's responsibility", it is more likely that it will be nobody's. As Stephen Elop, the former CEO of Nokia, complained, "We didn't do anything wrong, but somehow, we lost".

Drivers of Innovation Culture

Changing a corporate culture, particularly in large organizations, is probably one of the toughest management challenges. Moreover, when an organization realizes that it is urgent to change, it is often too late. Potential change levers include leadership, vision and management support for innovation; human resources and formal knowledge-management policies and decision-making processes as well as more informal levers such as values, climate and sense of urgency.

Examples of potential corporate culture change initiatives include "walking the talk" at top management level and "putting your money where your mouth is," as well as celebrating early successes and identifying and highlighting innovation role models within the organization.

So What?

Organization can successfully manage innovation in a sustainable way only if they can create an environment where people and teams have the motivation and possibility to learn and experiment across corporate silos and in spite of strong corporate routines. An innovation strategy not supported by an innovation culture is like a car without wheels and fuel. Innovation managers must keep in mind the advice of the famous management guru Peter Drucker, "Culture eats strategy for breakfast".

3.3.3 Doing Completely New Things: Corporate Ventures

Putting in place the right culture and systems will help an organization manage innovation and deal with a changing corporate environment. But there are cases where the culture and systems which work for the organization as a whole might not be the right ones for developing specific innovation opportunities, in particular those related to radical innovations. In this case, it makes sense to consider creating a "special approach" for such corporate ventures (Birkinshaw 1997).

If you want to play jazz with a classical orchestra, it is probably easier to pick a few musicians and isolate them in a separate room, rather than trying to convince the whole orchestra to play completely new tunes.

Never Seen Before but Obviously Core

Corporate ventures relate to innovation opportunities that fit within the strategy of the firm but cannot easily be integrated within its existing organization. When the innovation implies "bold" high-risk but high-potential projects, when it involves completely new markets, value propositions or value chains, or when it mobilizes resources and time horizons which lie beyond the scope of existing business units, it is often better to manage the venture in a "fast-track" environment with a distinct culture and systems.

For example, the company Grundfos has defined the criteria for corporate ventures as combining "new to the world" performance features, significant (five to ten times) improvement in known features and significant (30–50%) expected reduction in cost.

Far Away, but Still Home

The objective of a corporate venture is to "have its cake and eat it". It is to do completely new things while keeping the benefit of being part of a large corporation. One of the key challenges when setting up such a dedicated structure is therefore to define the right "distance" from and the right interfaces with the main corporate organization. If the new venture is too close to and too embedded in corporate processes, it might suffer from organizational inertia, lack of entrepreneurial skills and a focus on traditional sources of innovation. But if it is too far away and too disconnected it might fail to capture synergies with corporate assets.

When Microsoft entered the gaming business with the new Xbox, it created a completely separate business unit in another city. In particular, it wanted to make sure that the new venture could escape from the Windows/Microsoft Office culture and reputation and attract hard-core gamers and experts. But at the same time it made sure that having access to the corporate assets of Microsoft would give the Xbox a competitive advantage over Sony or Nintendo.

Designing Corporate Ventures

Corporations have experimented with multiple ways to manage corporate ventures "at the right distance", trying to maximize the probability of such projects becoming successful businesses and contributing to corporate objectives. These might involve totally embedded projects or completely new divisions reporting to the top management, opportunities closely related or completely unrelated to the core business, and projects sponsored "top-down" by management or employee-driven initiatives.

In all cases, corporations must consider when and how the venture will escape and then reintegrate into the "normal" organization, who will be the dedicated entrepreneurial team in charge of successfully scaling up the venture, and what will be the governance of the venture and of its interfaces with the corporation.

Examples of corporate venturing structures include dedicated business units and teams, joint ventures and spin-offs as well as projects managed by a new venture division and board, through a direct minority investment or at arm's length by a third-party incubator or "accelerator".

Parenting Corporate Ventures

For corporate ventures the challenge is to find the governance structure and processes that allow the new venture to leverage corporate economies of scale and learning curves and foster learning and synergies while at the same time be flexible enough to deal with its specific needs, skills and managerial styles. In particular, this involves defining the right interfaces between the new venture and both the corporate operations (facilities, customer base, brand, etc.) and its other innovation initiatives (R&D, marketing, M&A, etc.).

From the first accelerators of the Internet boom to the new incubators of industry 4.0, examples abound of corporate ventures that failed to take off (become profitable) or land (become a "normal" business in the corporate portfolio and have a significant impact on its bottom line). Too often corporate ventures neither eat their cake nor have it.

So What?

Managers should consider designing dedicated governance processes and structures when dealing with innovation opportunities that are within the scope of the strategy but outside the scope of the existing organization. But the key challenge will be to choose the right distance and interface between the corporate organization and the new venture governance. It must allow the new venture to both successfully take off and land, shed the disadvantages of being part of an existing firm and capture the benefits of being a fully independent entity.

3.3.4 "Ambidextrous" Organizations: Small Is Beautiful, Big Is Powerful

Organizations that successfully cultivate and manage innovation must maintain their ability to both generate newness and achieve change, which are the two sides of the innovation coin. This "ambidexterity" (Tushman and O'Reilly 1996) is particularly a challenge as firms grow, transitioning from fire-fighting ("start-ups") management modes to more formal decision processes and responsibilities, then scaling up processes to become corporate giants, sometimes hiring thousands of people across hundreds of locations and units.

Conversely, there is a mythology of the small start-up in a garage single-handedly disrupting the incumbent dinosaur, although the majority of small businesses actually fail to be innovative and/or to achieve any significant success. But behind the myths there are indeed advantages and disadvantages linked with scale. The challenge is to become powerful while remaining beautiful.

Start-ups supposedly rule the innovation world. But how many innovative products have you bought in the last year that were sold by a small business? While some sectors such as news, advertising, books or the music industry have been disrupted, incumbents still rule most of the economic world.

Big Is Powerful

Although many small firms actually do not want to grow, increasing scale, reach and assets has long been recognized as linked with corporate success. From an innovation management point of view, large size brings the ability to mobilize innovation resources and complementary financing as well as sales and marketing capabilities that are out of reach for most small organizations.

Professional experts, extensive and diverse experiences, brand, credibility and customer base, cash, market power, influence and networks give large firms the opportunity to capture economies of scale and scope, shape their environments and in some cases spot and exploit unforeseen results and combinations.

Large consumer goods firms such as Nestlé, Coca-Cola, Danone or Unilever have long leveraged their brand recognition, financial resources, scale economies and product placement experience to successfully launch new products and enter new markets.

Creative Accumulation

Large firms also often have the opportunity to spread risks and fixed costs as well as share good practices and learning across a wide range of innovation initiatives. Their scale also allows them to sustain shocks and survive larger losses, deal with complex regulatory environments and develop in-house specialized capabilities. Finally, large firms can often recruit, train, acquire or poach talent more easily and rely on a wide pool of managerial company-specific skills.

As an illustration, Jeff Immelt (its former CEO), stressed that "GE's strength is not in breakthrough inventions but in turning $50m ideas into billion-dollar ideas".

Lost in Translation

The main disadvantage commonly linked to scale is the difficulty of maintaining an innovative corporate culture and continuing to foster corporate entrepreneurship. Large firms often perceive that they have "a lot to lose" and want to protect the integrity of their existing business assets. They therefore create managerial controls, internal boundaries (silos) and bureaucracies that can hinder freedom, flexibility and creativity and hence prevent innovation initiatives from emerging.

Their sheer size can also deter employee engagement, as the ability of a prospective corporate entrepreneur to "have a visible impact", share new knowledge and make change happen is constrained in a large structure, while the risks remain high.

Sony once had the products, brand, technology and content to rule the music world, but its organization struggled to manage the MP3 and streaming revolutions.

Small Is Beautiful

Another disadvantage of large firms in terms of innovation is their strategic commitments to existing knowledge ("invented here"), employees, suppliers, partners and customers, which can generate conflicts of interest, for example, when cannibalizing existing products with a new one. Small firms have much less to lose in terms of existing values or commitment, and are therefore freer to move.

General Motors put an electric car on the market in 1995 but soon canceled the project, among others because of the way it threatened its existing profitability, business and partners.

Ambidexterity: Strong and Agile

Trying to achieve the power and resources of large corporations while maintaining the culture and agility of start-ups can generate conflicting objectives, styles and resource trade-offs. "Ambidextrous" organizations combine the efficient and focused management of current operations ("running the bank") with an effective and flexible management of innovation projects ("changing the bank"). They find ways to strike the right balance between alignment and complexity, facts and opportunities, routines and breakthroughs, data and pictures, consistency and diversity, resources and people, returns and success rates.

The challenge of ambidextrous organization is to be as effective as a well-designed machine while at the same time be as adaptable as an evolved living organism.

"Contextual" Versus "Structural" Ambidexterity

Some innovative firms focus on culture and flexibility across their organization and foster "contextual" ambidexterity. They mobilize specific corporate support services and processes to push all their people and teams to engage in exploration and exploitation over time and across projects.

Professional service firms such as the leading strategic consultants have developed and implemented horizontal governance processes that allow them to be locally reactive while remaining globally effective.

Other firms focus more on corporate ventures and acquisitions, specialization and "structural" ambidexterity. They acquire small businesses and divide their corporate structure into entities more focused on exploitation activities, such as a manufacturing department or a commodity business unit, and other entities more focused on exploration, such as a new business division, acquired start-ups or a niche business unit.

Conversely, small firms can gain power and become ambidextrous by partnering with or being acquired by large corporations, while maintaining sufficient autonomy to remain agile.

Large industrial firms such as General Electric or Siemens have a wide business portfolio that combines exploitation- and exploration-minded businesses.

Managing Ambidexterity

There are no unique ways for corporations to achieve such "ambidexterity". The right way for a given firm to achieve ambidexterity will be driven on the one hand by the strategic importance of corporate resources (synergies) and on the other hand by the complexity and dynamics of the corporate environment (agility).

Google is fostering an innovation culture across its whole organization but also creating dedicated entities for its "moonshot" projects. It is also acquiring and scaling up successful start-ups such as Waze. Conversely, the main exit for innovative start-ups is trade sales (being acquired by a big firm) rather than IPOs.

So What

From an innovation management perspective, size makes it more difficult to explore newness (scale is challenge), but it also makes it easier to make change happen (scale is an asset). The challenge for innovation-ready organizations is therefore to build an organization that combines the strength of large firms with the agility of small ones, by developing an innovation culture across the organization but also creating dedicated environments for innovative ventures.

3.4   Develop Innovative Networks and Collaborations: Never Walk Alone

Managers have long noticed that whatever the size of their firm, there would always be more skills and expertise available outside their organization than inside. There is therefore an opportunity to better manage innovation, with partners, by pooling assets and acquiring new capabilities. However, this requires developing new governance approaches, particularly regarding the firm's ability to identify potential partners and implement effective ways to collaborate with them (Fig. 3.5).

Key Insights

  i. Partnerships offer opportunities to capture unique competitive advantages, by gaining scale and speed—pooling resources—and by developing unique assets—acquiring new capabilities.

  ii. Building closer ties, particularly across industries, requires dealing with physical as well as cultural distance and being able to leverage innovation intermediaries, clusters and communities.

  iii. Open innovation is about systematically reaching out in order to better identify, develop and/or implement innovation opportunities with partners and outsiders.

  iv. Effectively managing innovation across corporate boundaries requires dedicated skills and capabilities, particularly regarding the identification and selection of the right partners and the design and implementation of the right partnerships.

Fig. 3.5 Developing innovation networks

3.4.1 Capturing Unique Competitive Advantages from Partnerships

Successfully managing innovations means identifying and capturing new opportunities better than others. While many firms have developed unique capabilities to do so in their specific markets or industries, it is very likely that useful skills and capabilities still exist outside those firms. This is particularly the case as more and more innovations actually cross traditional industry boundaries and therefore require firms to mobilize skill sets that go beyond what most of them can master.

This creates opportunities but also threats, because while a firm might choose not to open up, existing or new competitors might choose to do so and could gain significant competitive advantage.

As Bill Joy, the co-founder of Sun Microsystem, noted many years ago, "No matter who you are, most of the smartest people work for someone else."

Unity Makes Strength

The first way to gain competitive advantage through partnerships is by pooling resources and assets in order to capture synergies. This allows collaborating firms to achieve economies of scale. Such partnerships allow firms to save design and production costs; share learning, ideas and data; minimize risks; jointly shape new regulation and standards; and create lock-in and signal long-term commitments.

When Renault successfully launched its "Espace" family car, its main competitors in Europe chose to team up in order to save costs and minimize risks and to more quickly bring alternative models to the market. Similarly, European car manufacturers have been joining forces to catch up with Tesla and try to beat it at its own game. Finally, many engineering standards are defined based on the strengths and priorities of the consortia supporting them as much as on their intrinsic qualities.

Unique Combinations

The second way to gain competitive advantage through partnerships is by accessing complementary resources, in order to differentiate by building unique combinations. Collaborating firms can combine unique sets of technology, expertise, assets and intellectual property to capture existing and new opportunities. They can combine unique brand, reputation, market knowledge and reach to capture new markets. Finally, they can secure exclusive partnerships to create unique bundles of products and services.

The battle for the electric battery market will be won by the firms that orchestrate the best combinations of access to key raw materials (such as rare-earth elements), chemistry and material science expertise (membranes, electrolytes, etc.), design and manufacturing skills (weight, cost, reliability, recycling, etc.) and market knowledge (specifications, distribution, installation and maintenance, etc.). Similarly, the success of Nespresso has been based on a unique combination of procurement, manufacturing, design and retail skills.

So What?

Managers who operate in innovation-intensive environments should constantly scout for partnership opportunities within and outside their industry and identify opportunities to develop competitive advantages by pooling resources (synergies) or acquiring assets (unique combinations). If they do not, their competitors will, and preempt unique partnership opportunities. In most markets good partners are scarce and therefore valuable.

3.4.2 Building Closer Ties: Innovation Networks, Communities and Intermediaries

Finding and working with third parties can be challenging, particularly when dealing with risky and/or ambiguous innovation projects. Whether we like them or not, there are good reasons we have organizations and colleagues, with whom collaboration is supposed to be easier than with complete strangers. It is therefore important for firms in innovation-intensive environments to develop "weak" and "strong" ties with other organizations (Granovetter 1973), in order to identify potential partners and get "closer" to them.

For centuries formal and informal business networks have relied on trust, shared culture, language and values. When they do not exist a priori they need to be developed.

Dealing with Cultural Distance

While in the past firms collaborated mostly in tightly knit networks of peers and local communities, with "people they knew", many of them today have to operate in global industries and markets, spanning multiple disciplines, cultures, continents and time zones. The opening of the innovation process has also led businesses interact more with completely different actors, such as start-ups, freelance experts and SMEs, universities and research centers or public organizations and NGOs. Instead of the "death of distance" Internet gurus dreamed about, firms therefore increasingly need to find ways to cope with "cultural distance". They need to find smart ways to deal with differences not only in location and time zone but also in norms, routines, languages and processes related to the management of innovation.

Collaborations with "close" partners (in terms of cultural distance) can often be dealt with in an ongoing and relatively tacit way. But collaborations with far away "aliens" require much more careful and explicit management of the respective expertise and deliverables.

Many "fintech" innovation collaborations between "geeky" Internet champions and "traditional" financial institutions have failed not because they could not identify potential joint business opportunities, but because they could not find ways to work effectively together and reconcile their respective cultures, values and norms.

Networking: Building Closer Ties

One way to cope with cultural distance and find potential partners is to build weak and strong ties with other organizations through networking initiatives. Those innovation networks can be part of regional initiatives, "clusters" (Porter 2000) and innovation systems or include communities of international partners. They can address generic innovation issues or focus on specialized sectors (e.g. pharmaceutical or energy), topics (e.g. sustainability or intellectual property) or technologies (e.g. 3D printing or machine learning). They are particularly useful in dealing with the growing numbers of cross-industry innovations, where collaboration opportunities extend beyond traditional partners, customers or suppliers.

Technology clusters such as Silicon Valley and its dozens of copycats are examples of networks that combine regional proximity and technology focus. They facilitate interactions and trust between large firms, SMEs, universities, investors, markets and experts. While such clusters are difficult to institutionalize "top-down", they can, when successful, become a major source of innovations.

Innovation Communities: Sharing Ideas and Good Practices

At a more individual level, managers involved in innovation-intensive environments can also join innovation communities. In such communities, individual managers can on the one hand share ideas and inspirations (find new things) and on the other hand share good practices and build trust (do new things) with people from other organizations. These online and offline innovation communities can be highly fluid or involve tightly controlled membership and access. They can also have broad and informal purposes or involve tightly defined contributions and deliverables.

The GRD (Groupe Recherche-Développement) is a Benelux-based community of innovation and R&D managers launched in 1966. Since then hundreds of managers have invited one another to their respective facilities to meet for half a day every month and share good practices, build trust and nurture future cross-industry collaborations.

Innovation Dating Agencies and Facilitators

Like lonely hearts looking for the perfect partners, isolated firms can also sometimes rely on dedicated intermediaries to help them find potential innovation partners. These "innomediaries" can help them articulate what they are looking for, identify and screen potential candidates, and facilitate contacts and align expectations.

Innovation "community managers" can also play a role in the selection of members, allocation and regulation of roles and responsibilities and follow-up of activities and results.

But relying on such third parties can also create challenges in terms of competitive intelligence (competitors might find out what a firm is exploring), management of intellectual property (who owns and shares what) as well as social competencies and technical skills (does the intermediary really know what a firm is looking for?).

While publicly funded "innomediaries" have existed for decades in Europe (through what has become the Enterprise Europe Network), private initiatives include NineSigma, Innovation Change, yet2.com and Innocentive.

So What?

Steve Jobs famously said that innovation was about "connecting the dots". Innovation managers must therefore make sure that they invest enough time and attention in targeted (from a geographic or technologic viewpoint) networking activities, both from an organizational and personal point of view. These initiatives must help them both find new potential dots to connect and reduce the complexity of effectively managing the resulting connections. In innovation, "who you know" is often as important as "what you know".

3.4.3 Open Innovation (Chesbrough 2003): Proudly Found Elsewhere

Innovation has traditionally been managed behind closed corporate doors, because of confidentiality issues, because it is often simpler to work with colleagues or simply because it has always been done that way. But since Joseph Schumpter's "creative destruction" we know that markets can often be more efficient than individual firms at creating and killing potential innovations.

Moreover, innovations often require new market and technology capabilities that are not available in-house. Finally, the whole process of turning an idea into a business might be too long and too complex to be managed by a single organization. Therefore, firms have tried for many years to open their innovation processes to address these issues.

Innovations as old as the first marine chronometer (1714) or the first Oxford English Dictionary (1858) were achieved only by mobilizing a wide range of expertise and people. Civilizations as old as those of the Greeks or of the Romans have known that trying to "invent everything here" was less effective than relying on external sources of knowledge.

Sourcing: Absorbing Good Ideas

The first way to open the innovation process is to stop relying only on opportunities identified internally and to start exploring ideas and experimentation beyond corporate walls. Even firms with huge R&D and marketing departments and thousands of employees cannot assume that they have a monopoly on good ideas. Universities, neighbors, partners, customers and suppliers are sources of ideas and opportunities that should be "absorbed" by businesses in a systematic way.

In more extreme cases, the whole "crowd" of brains and imaginations can be called on, particularly now that Internet access and social networks are ubiquitous. Of course, relying in such a way on other people's ideas requires a careful management of rewards, expectations, confidentiality and intellectual property.

Pharmaceutical firms have long relied on new molecules developed by universities, research centers and "biotech" firms to fill their innovation pipelines. Lego is famous for allowing its huge fan base to submit new designs using its famous bricks.

Development: Never Walk Alone

The second way to open the innovation process is to realize that a firm might not have access to the skills and capabilities required to successfully develop an innovation. Innovations often rely on new and/or very specialized skills for which better, cheaper, less risky and more quickly available sources might exist outside the organizations. This can involve simple subcontracting agreements but also complex international joint development programs. In such cases, the management of the resulting consortia (and the risks and complexity it generates) often becomes as critical as the management of the innovation itself.

IBM used to manage the development of its ICT innovations in a vertically integrated and mostly closed way, progressing from raw materials to chips and devices, computers, operating systems, software and applications. It now offers new services relying on a complex ecosystem of partners, including open-source providers such as Linux.

Go to Market: Don't Do It Yourself

The third way to open the innovation process is innovations whose time horizon or market falls beyond the scope of the firm's strategy and that might still be profitably implemented by others, rather than simply frozen or abandoned. This is particularly the case when an innovation implemented by a firm in a specific sector/market or for a specific type of application can be implemented outside that sector, market and/or application field by a partner, for example, through a joint venture or a profit-sharing agreement.

IRIS was an innovative company that developed advanced OCR (optical character recognition) software for scanning and archiving multiple paper-based data sources, such as customer invoices and contracts. It achieved global scale through partnership agreements with large OEMs such as Canon and HP, which integrated its software within their hardware.

Licensing: Selling Your Ideas

A commonly used way to let a third party bring an innovation developed in-house to market is through licensing agreements, the transfer of intellectual property rights to a third party. Licensing agreements often involve complex negotiations regarding the degree of exclusivity, scope and duration, reference metrics (sales, profit, milestones, etc.) and the corresponding remuneration (royalties, lump sum, cross licensing, etc.) as well as the roles and responsibilities in case of infringement. However, they can allow firms to reduce or eliminate production and distribution costs and risks, reach new markets or applications, establish de facto standards and gain bargaining power.

Nintendo's game console has been struggling to compete with the offers of Sony and Microsoft. But Nintendo found a cheap and fast way to enter the smartphone games market by licensing its famous Pokémon and Super Mario ranges.

So What?

Innovation managers should challenge natural corporate tendencies to try to do everything in-house. They should systematically seize opportunities to exploit other organizations' ideas, jointly develop innovations and/or let others bring them to market. A careful and targeted use of these levers can allow a firm to innovate faster and better.

In most markets, even the largest firms do not try to manage innovation completely in-house.

3.4.4 Managing Innovation Across Corporate Boundaries

The promise of open innovation is the access to the nearly unlimited ideas, skills and talent available in open markets. But there are reasons why we have organizations and colleagues rather than only individual economic agents trading with each other. Working within an organization, with identified colleagues working under existing contracts, allows innovation managers to save significant time and so-called transaction costs (Williamson 1981). Working effectively outside an organization therefore means being able to develop the skills and capabilities to deal with new challenges and risks, which were unknown as long as everything was managed internally.

Many senior corporate managers who join start-ups discover the pain and discomfort of no longer having a large pool of colleagues to rely on. Outsourcing programs have long been known for the complexity, rigidity and costs they generate. Open innovation is too often about jointly addressing problems a firm wouldn't have if it worked on its own.

Dating: Search and Information Costs

When dealing with a colleague or another department, a manager can usually find out reasonably quickly who they are, what they can and cannot do and under what conditions. But when looking for a third party, you cannot just "google" good potential partners. Successfully managing open innovations therefore first requires building the capabilities to identify and assess potentially attractive partners.

Identifying potential partners means promoting the firm as an attractive partner (pull) and/or proactively scouting for target partners (push). This means being able to build a strong corporate image and reputation as an attractive innovation partner, promote its unique skills and create online and offline meeting and matching opportunities. It also means dealing with the flow of requests this might generate, prioritizing needs and managing the interfaces between prospective partners and in-house contacts.

Large firms in the ICT world or world-class research institutions such as the MIT are bombarded with collaboration proposals from all over the world and have to find effective ways to manage this flow. By contrast niche B2B industrial companies with limited public recognition might struggle to attract potential partners.

Screening and Sorting

Identifying good partners means screening and assessing potential candidates through careful due diligence processes. This means gathering and analyzing reliable information not only in terms of explicit performance (technologies, financial results, customer base, governance, etc.) but also in terms of more tacit dimensions such as people, style, values, reputation, reliability, ambitions and expectations. It also means understanding the potential competitive implications of making a deal or deciding not to, owing to existing strategic commitments and commercial relationships, potential links with competitors, risks of pre-emption, exclusivity, etc.

Large corporations exploring collaborations with start-ups find themselves confronted with a wide range of actors and characters who often have limited track records and fuzzy governance structures. They have to develop the connections and forensic capabilities needed to screen hundreds of potential candidates and find the few needles in the haystacks.

Engaging: Bargaining Costs and Conflict Resolution Mechanisms

When dealing with internal resources, the terms and conditions of a collaboration are mostly defined using existing contracts, corporate processes, job profiles, service-level agreements or explicit missions. But when dealing with a third party, particularly for the first time, whether they are interested in collaborating and under what terms have to be defined and agreed on.

Successfully managing open innovation therefore also requires building the capabilities to identify and negotiate potential win-win partnership agreements. And designing and negotiating successful innovation agreements is an art few managers actually master. It requires jointly defining carefully the negotiation process (who, when, how, where, with access to which data), building trust and agreeing on the key dimensions and risks of a potential agreement. It also requires identifying the non-negotiable issues (if any) and finally finding ways to close and conclude the negotiation, either through exit or contracting. In particular, it means agreeing on the right conflict resolution mechanisms necessary to dealing with the intrinsic ambiguity and uncertainty of innovation projects.

Many partnership negotiations between corporations and start-ups fail not because there is no space for an agreement but either because parties have irreconcilable expectations or because the start-up is not ready to cope with the complexity and bureaucracy of many corporate contracts. The small firm might also simply be afraid to be overwhelmed by the size and sheer power of its prospective corporate partner.

Marrying: Trust, but Verify

Managing innovation projects with direct colleagues is already often a challenge, given the complexity, uncertainty and pressure many of those projects have to cope with. Managing and integrating such projects with outside partners is even more complex. Partners will have to monitor and enforce along the way the agreed property rights, tasks and deliverables, particularly regarding their respective knowledge and competitive position. They will also have to adjust along the way and integrate the likely environmental changes and disruption, such as internal changes within a partner (particularly staff turnover), technology, market and competitive evolutions, project delays and unexpected (negative or positive) results.

Innovation managers negotiating open innovation partnerships should not forget Zug Ziglar's observation that "Many people spend more time in planning the wedding than they do in planning the marriage".

Living Together

Innovation partners must find ways to combine their potentially conflicting systems, incentive processes, paces, cultures and styles while maintaining management attention and commitment. They also must one day be able to close the partnerships, discuss potential follow-ups and make sure the knowledge created is not wasted.

One key trade-off in such cases relates to the flexibility of the internal innovation management processes. If they are too rigid, some partners might be unable or unwilling to adopt them. But if they are too flexible and adjusted for each open innovation project, they might become too complex to be managed efficiently.

An important implication of all these challenges is that firms often prefer to stick with the same partners. This allows them to build trust over time, align incentives (the relationship matters more than a single project), develop framework agreements and learn to work together.

Partnerships between corporations and universities are notorious for the culture shocks and conflicts they can generate, given the different pace, values and incentives of these two types of institutions. The most successful ones are often those which can leverage long-term and personal trust relationships, both at operational and executive levels.

So What?

Open innovation offers great opportunities for innovation management but it also generates significant challenges. Effectively managing innovation across corporate boundaries therefore requires firms to develop dedicated skills and processes in terms of partner identification, due diligence, negotiation and collaborative project management. These are whole sets of skills that are often not found inside organizations and need therefore to be carefully developed.

3.5   Create Innovation Ecosystems: Lands of Opportunities

Since the dawn of civilization, some places have been hotbeds of knowledge creation and innovation while others have fallen into oblivion. More recently, regions across the world have tried to emulate the successes of entrepreneurial ecosystems such as Silicon Valley.

It is therefore critical for policymakers in particular and for the citizens who elect them in general to understand how and why the jobs and wealth of tomorrow can be created through new and growing ventures. The alternative is to wait and see the industries of yesterday disappear.

Conversely, innovation champions must learn how to understand the strengths and weaknesses of the innovation ecosystems in which they operate, and leverage the capacity of those ecosystems to foster their entrepreneurial ventures (Fig. 3.6).

Key Insights

i. The strength of a regional innovation ecosystem is driven by the combination of effective infrastructures and institutions with the availability of relevant financial, human and knowledge resources.

ii. Beyond the start-up myths, weaknesses such as lack of talent and ambition imply that most new firms emerging in a regional ecosystem will be low-growth and low-tech.

iii. Untamed free markets often fail to support sustainable innovations. Targeted and effective public interventions are also needed for strong innovation ecosystems to emerge and strive.

iv. A wide range of private and public innovation support mechanisms should be carefully deployed and leveraged in order to strengthen regional innovation ecosystems and foster the scale-up of entrepreneurial ventures.

Fig. 3.6 Creating innovation ecosystems

3.5.1 Innovation Valleys: Regional Innovation Ecosystems

Innovation does not "pop up" on its own and out of the blue. Innovation happens when new things are done in a specific context. The characteristics of that context as much as the characteristics of the innovation itself will influence the development of the innovation. It is therefore important for innovation managers to understand the key strengths and weaknesses of the regional innovation ecosystems in which they operate. Those characteristics will influence the emergence of their innovations, of innovation in general, and of new entrepreneurial ventures the firm could work with.

Bricks (Not Only Clicks) and Blue (Not Red) Tape

War zones or deserted islands might in some cases foster creativity and bricolage, but they are poor ecosystems for innovation. Innovators need institutions and policies to allow them to operate and in particular protect their property rights. Innovators need standards, regulations and infrastructures as well as virtual and physical spaces to compete and support their development. Finally, they need communication and logistical means to reach and interact with potential partners, customers and investors.

Even a virtual start-up that exists in the cloud ultimately relies on the availability of a telecommunication infrastructure and a payment system, as well as rules and institutions that protect its physical and intellectual assets. It would probably struggle in an environment with unreliable infrastructure and/or corrupt institutions.

Dots to Connect

Innovations are also built on new combinations of existing people, ideas and tangible resources, and are therefore constrained by their local availability. Innovation resources include the people with the right skills, culture and social structures to inspire, build and lead innovative teams and organizations. They also include formal and informal networks of potential customers, partners and suppliers—small and large—that are ready to do business with innovators. Finally, they include local research and development capabilities, in particular their technological sophistication and specialized expertise.

Some cities have become the new "places to be" for innovators. The urban innovation ecosystems of London, Tel Aviv, Barcelona or Berlin indeed provide easy access to a wide range of technical, financial and commercial skills and resources.

So What?

Innovation managers must understand and manage the regional innovation ecosystems in which they operate or plan to operate. The strengths and weaknesses of regional innovation ecosystems will strongly affect the development of innovations, particularly in emerging countries with weak infrastructures and unstable institutions. But this diversity also provides innovation managers with opportunities to cherry-pick their investment locations, as regional ecosystems compete to attract them.

Cities have been fighting against each other to be "the" ecosystem where self-driving cars will be developed. Similar institutional competition exists for drones, biotechnologies and artificial intelligence applications.

3.5.2 Muppets and Unicorns: Start-Up Myths

Popular magazines and websites are full of stories of start-ups disrupting whole industries single-handedly and overnight. Many regions have therefore engaged in broad and costly programs to foster new business creation, hoping to generate the next unicorns and hence create numerous new employment opportunities. Similarly, many large corporations have launched employee-driven or open "incubation" programs, hoping to harbor the technology champions of tomorrow. But managing innovation ecosystems by only listening to such stories is like designing a lottery system by only listening to the claims of the winners.

The Start-Up Mirage (Shane 2009)

The truth is not only that many new firms fail, but also that most of the surviving ones fail to grow and generate any significant economic impact. Similarly, most employee-driven ideas are unrealistic or irrelevant for their corporation, and most of the few remaining ones will ultimately fail to have any visible impact on the corporate bottom line. The road to entrepreneurial success is much more difficult than most policymakers or corporate managers imagined when launching their initiatives.

Google, Uber or Airbnb took many years to gain recognition and exert impact. The ratio of successful IPOs per idea in Silicon Valley has been estimated to be around one out of six million. Similarly, only 0.6% of the new businesses created in the United States during the last ten years can be considered high-growth "gazelles" (according to the Kaufmann Foundation). Banks are full of internal "start-ups" that have yet to impact the financial results featured in their annual report.

The Typical "Start-Up": A Low-Tech "Muppet", Not a Unicorn

Most of the new businesses created in a given regional or corporate ecosystem will employ no one other than the founder, and most of the remaining ones will only hire a handful of employees. Most new firms remain home-based, are less productive than their older peers, fail to innovate and are dead within five years. Among surviving ventures, only a small fraction will ever grow beyond micro-firms in terms of sales, assets or number of employees.

Finally, among the survivors, successful (high-growth) start-ups are not more prevalent in hi-tech industries than within low-tech industries. In a nutshell most new firms are what Paul Nightingale called "M.U.P.P.E.T.S.": Marginal Undersized, Poor Performance EnTerpriSes.

Successful innovative start-ups are not only rare but also dispersed. Most start-up rankings based on rigorous statistical sampling rather than individual perceptions, surveys or Internet buzz identify successful high-growth ventures also emerging from supposedly "low-tech" sectors or industries, such as retail or professional services.

Why They Do Not Grow

When the managers of small businesses are interviewed, they often complain about what annoys them or say how they think they can be helped: bureaucracy, taxes and lack of funding. But someone who thinks that filling an administrative form is an unbearable hurdle has probably never sold a complex product. And regions with similar tax systems and funding opportunities often show widely different results in terms of innovation and entrepreneurship.

While access to resources can be an obvious hurdle, the main reason why many promising small businesses do not grow is simply because they do not want to. Many small business managers do not see growth as an objective per se, and certainly even fewer aim for growth in terms of employment.

Most of the new businesses created by generic public or corporate policies will therefore fail to deliver the jobs or profits their patrons dream of. And in terms of scarce resources, the biggest problem for ambitious (corporate or stand-alone) entrepreneurs with attractive opportunities is people, not money. They can find funding, subsidies or investors, but they cannot find the leaders, team members and partners they need.

The biggest hurdle for firms that have to deal with a digital transformation is not a lack of new start-up ideas or (in most cases) a lack of money. It is a lack of people with the will and skills to capture new business opportunities. The biggest ally of a new business development manager should be the human resource manager, not only the CFO or the CTO.

So What?

Public or private managers who try to foster innovation through the development of entrepreneurial ecosystems and start-ups need to focus on quality rather than quantity. They need to select projects not only based on potential market or significant competitive advantage but also on the availability of teams with the right skills and ambitions. It is therefore often more productive to help a single existing successful venture to upgrade and grow than to try to create 20 new ones.

3.5.3 When Markets Fail: Targeted and Effective Public Interventions

On top of the policies aimed at fostering new business creation, managers and economists have long recognized that targeted and effective public interventions might be needed for innovations to flourish. The "invisible hand" of free markets does not always guarantee the best allocation of resources. In the *Wealth of Nations*, Adam Smith himself supported policies aimed at encouraging inventions and new ideas through patent enforcement and support of infant industry monopolies. Targeted and effective public interventions are needed on one hand to create and shape new markets and on the other hand to ensure the efficient functioning of those markets. Strong innovation ecosystems require strong governance.

While public (and corporate) money has sometimes been wasted on the pursuit of "white elephants", major innovations such as satellites or biotechnologies could not have emerged without targeted support and subsidization. Larry Page, Sergey Brin and Mark Zuckerberg are certainly very bright and hard-working, but without the Internet they would probably be nowhere, and the Internet was created by public money.

Creating and Shaping New Markets

For the invisible hand to work its magic, it first needs a market where goods can be supplied and demanded. The creation and shaping of such markets requires public policies and regulations, to ensure property rights and contract enforcement on one hand and to create a level-playing field and ensure fair competition on the other hand. It also often requires long-term investments in early stage research and development, for technologies with potential general purposes but no obvious short-term applications. In sum, you need a field and a referee before entrepreneurial players can score goals.

The laser was for many years a "laboratory gadget" with no known applications. Without long-term public support and standardization policies it would probably have remained so and not become the ubiquitous innovation we now all benefit from. Solar panel technologies could never have emerged as an alternative energy source if they were initially left to freely compete with incumbent solutions.

Selling Apple Versus Selling Lemons (Akerlof 1970)

Once a market is in place for a given innovation, it will work efficiently only if buyers and sellers can agree on a good, a quantity and a price. But most innovations cannot be traded like simple commodities. They embed knowledge that cannot easily be sold. They involve tacit knowledge and information asymmetries that make it difficult to clearly define which "good" or service is actually traded. They often require upfront investments, generate uncertainty and are hindered by risk aversion. Finally, some customers might need the innovation but are unable to access or pay for it. In sum, there is a need for many "visible" hands in order to allow the supply and demand of innovations to meet and trade.

New pharmaceutical drugs are examples of goods for which the regulation of intellectual property, quality controls, infrastructures, customer information campaigns and public assistance for insolvent customers are often needed for the resulting market to effectively maximize welfare. As a consequence, public authorities and private foundations, such as the Bill and Melinda Gates Foundation, have long played an important role.

Managing Side-Effects

An innovation market will flourish and maximize welfare if it also integrates the externalities it generates. When A buys something new from B, it often positively or negatively affects the welfare of C. Innovations are not managed in a vacuum. The development of a given innovation will often positively or negatively affect the economic development of whole regions or industries. It can also generate economies of scope and knowledge spillovers.

Finally, innovation investments made today can irreversibly affect the technology paths of tomorrow. Innovations create wealth and welfare but also sometimes destroy jobs and threaten future generations. And the uncertainty surrounding their impact often creates fears that make things worse. There will be (perceived) winners and (perceived) losers for each innovation, and they should not be ignored.

Innovations related to new energy technologies or artificial intelligence opportunities should not be managed without considering their overall environmental and/or social impact. Similarly, technologies related to animal testing or genetically modified goods can generate strong social reactions. Since the Luddites, the English textile workers who protested against threatening new technologies in the nineteenth century, we have learned that such backlashes can be serious and have to be managed.

So What?

Entrepreneurship is the engine of innovation. But innovation managers need to recognize that effective public intervention is needed for most innovations to flourish. They need therefore to understand which interventions to foster regarding the creation and shaping of new markets, the effective trading of innovation-related goods or services on those markets, and the integration of negative and positive externalities, real or perceived.

Online intermediaries such as eBay, YouTube, Airbnb or Uber, or sustainable energy players such as Tesla, have long recognized that the way the markets in which they operate are and will be regulated will have a huge impact on the profitability or liabilities of their business models. The management of nuclear energy-related innovations has also been driven by their potential side-effects (actual and perceived) as much as by their intrinsic economic performance.

3.5.4 The Visible Hands: Innovation Support Mechanisms

On top of global organizational innovation levers such as culture, structure and governance, public and private innovation managers can put in place dedicated mechanisms aimed at supporting specific innovation processes. But the road toward such mechanisms is too often paved only with good intentions. Too often, the managers fail to practice what they preach to innovators: define an objective, identify the target segments and needs, design an offer and, finally, assess its impact.

Examples of public or corporate innovation support mechanisms with various levels of success include dedicated facilities (accelerators, incubators), events (award ceremonies, fairs, conferences, "hackathons"), coaching and support networks, competitions and prizes as well as privileged ("fast-track") access to funding or other resources.

Who Do You Want to Help?

The first step when designing or using an innovation support mechanism should be to define the objectives and the implications in terms of target audience. People-driven initiatives aimed at fostering engagement and awareness should be inclusive and target a wide audience. But venture-driven initiatives aimed at generating significant economic value should be selective and support only a small number of high-impact projects. Similarly, people-driven initiatives can start early with potential entrepreneurs generating potential ideas, while venture-driven initiatives are often more effective when they support more mature teams and projects to help them scale up. Fostering ideation is one thing; accelerating business scale-up is another.

Profit-driven business development players such as venture capitalists support less than 1% of all businesses created in a given year, and tend to prefer more mature projects with proven business models. Even then, only a fraction of the ones they support will ever succeed. But the few winners can have a disproportionate innovation impact.

What Is the Help They Need?

The best way to help people innovate will be a function of the industry, technology, geography, market, local culture, education, language and the innovation target audience (potential entrepreneurs or emerging ventures). Potential entrepreneurs can be encouraged by role models to increase their need for achievement and tolerance for ambiguity or to change their attitude, perception and intention in terms of entrepreneurship. On the other hand, emerging ventures can be influenced by policies and tools, fostering their technical and managerial skills, their motivation and networks and finally the strengths of their business models.

The best innovation incubators tend to focus on a specific industry and project maturity. They are also highly selective and provide high-quality services to a small number of high-potential projects. The worst innovation incubators end up trying to do everything for everybody and in the end only maximize their occupancy rates or the wishes of their corporate or political patrons.

How Do You Want to Help?

Since the apprenticeships of medieval craftsmen, many approaches have been used to support the development of new ventures. They include financial support provided directly through loans, subsidies or equity or indirectly through guarantees, incentives, contacts or recommendations. They also include logistical support such as flexible and affordable infrastructures and facilities, privileged access to business services or networking initiatives. Finally, they can include advice and training through peer learning and communities of practice, direct teaching or coaching (generic or specialized) or one-stop contact points for information. In some cases innovation support mechanisms can also include demand-side support through favorable regulations, labels, or privileged access to public or corporate procurement for new firms.

As an example, the Y Combinator has provided a combination of funding, networking and advice to a very select group of start-ups (their acceptance rate is lower than Harvard's) including the likes of Airbnb and Dropbox.

Deliver Efficiently

The efficient management of an innovation support mechanism goes beyond applying the best practices in professional services or infrastructure management. One of the key challenges is that these structures often involve multiple stakeholders with often conflicting objectives. First, the corporate or public patrons might have public relations (e.g. to be re-elected) or strategic objectives (e.g. in terms of geographic scope) that conflict with the objectives of the new ventures. Second, many entrepreneurs by nature do not like to be told what they should do, or have unfavorable perceptions of the people trying to help them. They therefore do not recognize that they might benefit from the available support, or do not find the support they like.

In some cases entrepreneurs can also milk the provided services or opportunistically exploit them. The impact of the provided support is indeed often difficult to measure objectively, given the absence of control groups (what would have happened without the support?), the role of multiple factors (luck, external shocks) and the delayed effect.

The innovation support infrastructures of many regions look more like complex "lasagnes" of multiple initiatives added over time, with unclear and changing objectives as well as often overlapping scopes. These regions now even need special advisors to help innovators find the right support services.

So What?

Innovation managers who want to strengthen their innovation ecosystem and develop or leverage dedicated innovation support mechanisms must make sure that they carefully align the services offered with the target objectives and needs. They also need to put in place targeted qualitative and quantitative indicators, in order to monitor the quality and impact of the services provided. High impact can sometimes be achieved, but only with patience and through careful and very selective design.

3.6   Synthesis

Manage Entrepreneurial Ecosystems: Key Insights

3.1. Encourage people to innovate: corporate entrepreneurs

i.   Innovation is made by people. But most innovators are not natural born entrepreneurs. They are not lone heroes creating single-handedly revolutionary products. They are ordinary people, often within existing firms, who decide to change things and are ready to cope with the resulting uncertainties.

ii. Corporate entrepreneurs do not plan, do, check and act like traditional managers. They focus on finding and mobilizing available resources in order to identify, assess and launch new initiatives, while adjusting goals and expectations along the way.

iii. Entrepreneurial intentions and behaviors can be managed and fostered. Tolerance for failure, reward systems, role models and norms, as well as skill development and slack resources can create the right attitudes, norms and perceptions.

3.2. Build and lead effective innovation teams: balancing acts

i.   Organizations can successfully manage innovation only if they have the ability to build and lead effective project teams around their innovation initiatives.

ii. An effective project team will become a great innovation team if it can manage key trade-offs in terms of level of pressure, diversity, slack and tolerance for diverging and outside perspectives.

iii. Building an effective innovation team means mobilizing transformational leaders and team members with the right profiles, background, motivation and skills—not just whoever happens to be "on the bench" when the initiative is launched.

3.3. Build and manage innovation-ready organizations: how some elephants can dance

i. Innovation-ready organizations are not innovative for the sake of it. They combine operational efficiency with both the corporate culture required to continuously exploit and do things in new ways—innovative mindset—and the corporate structures required to explore and sometimes do completely new things—innovative ventures.
ii. Innovative organizations manage to develop a corporate culture that fosters trust, learning and exchanges in order to do things in new ways and over-come the traditional organizational barriers to corporate entrepreneurship.
iii. Innovative organizations can create dedicated structures where completely new corporate ventures can be nurtured and parented, leveraging their cor-porate assets while freeing "teenage" ventures from organizational inertia.
iv. Ambidextrous organizations are designed and managed to retain the inno-vation agility of many small firms, such as flexibility, engagement and auton-omy, while capturing the managerial efficiency of large corporations with their scale, assets and power.

3.4. Develop innovative networks and collaborations: never walk alone

i. Partnerships offer opportunities to capture unique competitive advantages, by gaining scale and speed—pooling resources—and by developing unique assets—acquiring new capabilities.
ii. Building closer ties, particularly across industries, requires dealing with phys-ical as well as cultural distance and being able to leverage innovation inter-mediaries, clusters and communities.
iii. Open innovation is about systematically reaching out in order to better iden-tify, develop and/or implement innovation opportunities with partners and outsiders.
iv. Effectively managing innovation across corporate boundaries requires dedi-cated skills and capabilities, particularly regarding the identification and selection of the right partners and the design and implementation of the right partnerships.

3.5. Create innovation ecosystems: lands of opportunities

i. The strength of a regional innovation ecosystem is driven by the combina-tion of effective infrastructures and institutions with the availability of rele-vant financial, human and knowledge resources.
ii. Beyond the start-ups myths, weaknesses such as lack of talent and ambition imply that most new firms emerging in a regional ecosystem will be low-growth and low-tech.
iii. Untamed free markets often fail to support sustainable innovations. Targeted and effective public interventions are also needed for strong inno-vation ecosystems to emerge and strive.
iv. A wide range of private and public innovation support mechanisms should be carefully deployed and leveraged in order to strengthen regional innova-tion ecosystems and foster the scale-up of entrepreneurial ventures.

Bibliography[1]

Akerlof, G. A. (1970). The market for "lemons": Quality uncertainty and the market mechanism. *The Quarterly Journal of Economics, 84,* 488–500.

Amabile, T. M., Conti, R., Coon, H., Lazenby, J., & Herron, M. (1996). Assessing the work environment for creativity. *Academy of Management Journal, 39*(5), 1154–1184.

Baker, T., & Reed, E. N. (2005). Creating something from nothing: Resource construction through entrepreneurial bricolage. *Administrative Science Quarterly, 50,* 329–366.

Bass, B. M. (1990). From transactional to transformational leadership: Learning to share the vision. *Organizational Dynamics, 18*(3), 19–31.

Birkinshaw, J. (1997). Entrepreneurship in multinational corporations: The characteristics of subsidiary initiatives. *Strategic Management Journal, 18,* 207–229.

Burgelman, R. A. (1983). Corporate entrepreneurship and strategic management: Insights from a process study. *Management Science, 29*(12), 1349–1364.

Chesbrough, H. (2003). The governance and performance of Xerox's technology spin-off companies. *Research Policy, 32*(3), 403–421.

Granovetter, M. S. (1973). The strength of weak ties. *American Journal of Sociology, 78*(6), 1360–1380.

Kogut, B., & Singh, H. (1988). The effect of national culture on the choice of entry mode. *Journal of International Business Studies, 19*(3), 411–432.

Kor, Y. Y., & Mahoney, J. T. (2004). Edith Penrose's (1959) contributions to the resource-based view of strategic management. *Journal of Management Studies, 41,* 183–191.

Nohria, N., & Gulati, R. (1996). Is slack good or bad for innovation? *Academy of Management Journal, 39*(5), 1245–1264.

Porter, M. E. (2000). Location, competition and economic development: Local clusters in a global economy. *Economic Development Quarterly, 14*(1), 15–34.

Shane, S. (2009). Why encouraging more people to become entrepreneurs is bad public policy. *Small Business Economics, 33*(2), 141–149.

Tushman, M. L., & O'Reilly, C. A. (1996). Ambidextrous organizations: Managing evolutionary and revolutionary changes. *California Management Review, 38*(4), 8–31.

West, M. A. (2002). Sparkling fountains or stagnant pounds. *Applied Psychology: An International Review, 5*(3), 355–524.

Williamson, O. E. (1981). The economics of organization: The transaction cost approach. *American Journal of Sociology, 87*(3), 548–577.

---

[1] An extended bibliography is available at www.NavigatingInnovation.org

4

Identify Attractive Innovation Opportunities

Innovation opportunities do not magically "pop up" out of the blue, be it in R&D laboratories or brainstorming sessions. They emerge when organizations "learn to learn" from multiple internal and external sources, increasing their knowledge and intellectual capital by "thinking in new boxes", mixing, maturing and combining multiple insights and inspirations.

The third innovation management challenge is therefore to effectively identify innovation opportunities, by developing the capabilities to systematically develop, screen, protect and combine both organizational and external sources of innovations (Fig. 4.1).

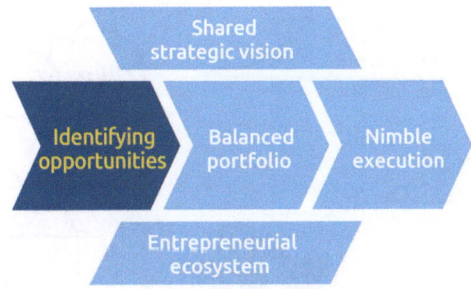

4.1 Identify the sources of innovation opportunities
4.2 Foster organizational learning: beyond ideation
4.3 Harvest and protect organizational knowledge assets
4.4 Integrate external sources of knowledge
4.5 Synthesis

Fig. 4.1 Identifying attractive innovation opportunities

© The Author(s) 2018
B. Gailly, *Navigating Innovation*, https://doi.org/10.1007/978-3-319-77191-5_4

4.1 Identify the Sources of Innovations: Beyond R&D

Identifying innovation opportunities better and faster than competitors requires more than outspending them in R&D. It requires understanding first where the new ideas leading to those opportunities come from, inside and outside the organization (Fig. 4.2), and the implications of trying to combine and mature those new ideas in order to generate actual business opportunities.

Key Insights

i. *R&D is not enough.* Significant R&D spending might be necessary for some firms to develop new technologies but certainly do not guarantee their innovation success. Identifying innovation opportunities requires finding new combinations of new or existing technologies with new or existing needs.

ii. The *triggers of innovation opportunities* which firms should proactively exploit include internal sources such as new knowledge, challenging routines and serendipity, and external sources such as changes in markets, industry and environmental constraints.

iii. *Sizeable innovation opportunities* do not pop up out of the blue. Organizations must invest time and resources to combine, integrate and mature innovation ideas into potential opportunities. They should also pursue already emerging opportunities and focus on scaling them up.

Fig. 4.2 Identifying the sources of innovation. (Adapted from Drucker 1985)

4.1.1 R&D Is Not Enough

Probably the most commonly used measure of innovativeness for a firm or a region is its level of R&D spending, often quantified as a share of total expenditures or revenues. As a consequence, firms and countries across the world continuously benchmark each other and policymakers regularly make frantic calls for more investment in R&D.

Causes and Effects

The link between levels of R&D spending and value creation is weak at best and might actually be the other way around (the wealthier you are, the more you invest in R&D). Multiple studies have failed to identify any discernible statistical relationship between R&D spending levels and nearly all measures of business success. Differences in R&D spending are actually often better explained for firms by differences in business models and for regions by differences in industry structures.

General Motors was one of the biggest R&D spenders just a few years before going nearly bankrupt and being rescued by the US government. From 2006 to 2016, Alphabet (Google's parent company) saw its return on invested capital fall just as its R&D spending increased from 10% to 15% of revenue. The level of R&D spending of most mid-size European countries is linked more strongly to the investment decisions of a handful of large multinational corporations than it is to their "intrinsic" innovativeness.

Combining Technologies and Needs

Effective R&D investments allow some firms to develop new technologies. But these new technologies are just new applications of knowledge, new ways to use resources, new combinations of assets and skills. They are relevant only if they perform value-creating tasks, by exploiting resources to meet somebody's need. Technologies are needed to solve existing and new problems, and new technologies create value only if they help some people solve some problems. All innovations need technologies—new or old—but not all technologies lead to innovations.

The so-called organizational, social or business model innovations all involve the use of some technologies. "Non-technology" innovation is somehow an oxymoron and may entail helping some "soft science" departments get their fair share of public subsidies related to innovation. All innovations mobilize technologies, but only some innovations are R&D-based.

Push and Pull

Innovation is therefore not only about new technology. Innovation is about both using new technologies and using existing technologies in new ways, to meet new or existing needs. Innovation opportunities can therefore be identified as either "inside-out", when new ways to use resources are "pushed" to address existing needs, and "outside-in", when needs are "pulled" to find new ways to satisfy them.

While "push" approaches might be very creative, they often generate significant waste: useless technologies are developed. "Pull" approaches are often more efficient but can be either shortsighted, as most customers tend to ask only for incremental rather than radical innovations, or rely upon unrealistic expectations ("free lunches").

New technologies that have initially struggled to meet a known market need include laser and carbon nanotubes. Unmet needs for which firms struggle to find distinctive solutions include obesity and time travel. Finally, successful (radical) innovations which were not initially perceived as "useful" by market research include disposable razors, personal computers and mobile telephony.

Another risk of relying too much on market-driven "pull" approaches is their potential lack of competitive differentiation, as most customers will often express the same (short term) needs to multiple competitors. Only firms with distinctive capabilities to understand and uncover untapped customer needs better than their competitors can afford to rely mainly on "pull" innovation approaches.

Large "FMCG" (fast-moving consumer goods) companies and retailers have developed sophisticated ways to find unmet needs, exploit cognitive biases and sometimes shape customer preferences, by combining psychology and neurosciences with "big data" and advanced analytics. They now understand how contextual advertising, shop architecture, music, smell and multiple other seemingly irrelevant factors actually significantly influence consumer behavior. As an example, "pronounceable ingredients" in food has been shown to be a key buying factor for some US consumers, meaning that "sugar" could be preferred as an ingredient to "ascorbic acid" (vitamin C).

So What?

Investing in new technologies is sometimes necessary but never enough for firms (or regions) to succeed. Managers must combine—for example, through rapid prototyping approaches—the disruptive potential of "push" approaches with the efficiency of "pull" approaches. This must allow them to identify innovation opportunities to use new technologies and to use existing technologies in new ways in order to address new or existing needs.

Engineers are from Mars, marketers are from Venus. Engineers might claim to have designed and built every important parts of modern civilization. But marketers have made sure many people actually used them. One of the biggest challenges (and most reliable success factors) of innovation therefore remains the ability of organizations to foster collaboration and cross-fertilization between, on the one hand, technology- and feature-driven engineers and scientists ("It works!") and, on the other hand, customer- and benefit-driven salesmen and marketers ("It sells!").

4.1.2 Triggers of Innovation Opportunities: Beyond New Tech

Innovations are triggered when new ideas and new combinations emerge. Understanding where such processes can start allows organizations to more systematically search for innovation opportunities across the board. Such a proactive approach increases the probability that they will identify more attractive innovation opportunities, more effectively and before their competitors do.

Innovations can be triggered by multiple types of events that occur within or outside an organization. "Internal" triggers include new knowledge ("Eureka"), challenging routines ("What if?") and serendipity ("Wait a minute…"). External triggers include changes in markets, industry and environmental constraints.

Eureka

"Eureka" triggers refer to discoveries made by members of an organization. They can relate to new technologies designed in laboratories, new scientific observations made in research centers or new patterns emerging from the analysis of large data sets. These triggers relate to the classic view of invention, when "the lightbulb lights up".

While probably much more famous than prevalent, discoveries remain an important trigger of innovation, particularly for radical innovations.

Classic examples of "eureka" triggers include the design of the first lightbulb by Thomas Edison after hundreds of trials and errors, and the discovery of the structure of human DNA by James Watson, Francis Crick and Rosalind Franklin.

What If?

"What if" triggers refer to when people question a routine or a process that is taken for granted in a given industry but is actually incongruous. These triggers relate to "positive deviants", those who manage to challenge dominant orthodoxies and the ways "things have always been done". While often strongly constrained by corporate routines and inertia, challenging routines is a particularly important trigger for process and service innovations.

Innovations triggered by "what if" thinking include fast-food restaurants, local currencies and "sharing economy" businesses such as Airbnb.

Wait a Minute…

"Wait a minute" triggers refer to new insights generated by accidental events and unexpected results. They include innovations unsuccessfully developed with a specific need in mind but that actually end up successfully meeting another need.

In a corporate setting, exploiting such innovation triggers requires not only the "luck" of generating unexpected positive results but also a culture and a governance that allow "failures" to be tolerated, shared across the organization and its silos and finally learned from.

From X-rays to Post-It notes (using a glue that does not really glue) and Aspartame (a powder with an unexpectedly sweet taste), such accidents are an integral part of the history of science. Less-known examples such as Zyban or Viagra relate to drugs developed for a specific disease, that generate unexpected side-effects which actually become their main source of value creation.

New Market Needs

New market needs relate to the classic "market pull" triggers, when an innovation is generated by a previously absent or unknown need being spotted among existing and/or potential customers. It can refer particularly to researching new trends and patterns both in customer demographics and in customer behaviors, preferences or perceptions.

Innovations triggered by market needs include incremental product improvements but also more radical and disruptive innovations. They represent a particular challenge for many B2B organizations, which increasingly have to better understand both the new needs of their customers and the needs of their customers' customers (consumers and end-users).

Changes in demographics that trigger innovations include population aging and migrations (e.g. the "gray" industry and new "expat" or "ethnic" services) as well as evolutions in customer levels of education or family structure (e.g. online education platforms and online dating services). Changes in preferences or perceptions include new priorities such as sustainability, fairness or economic patriotism ("Made in Here").

Changes in Industry

Changes in industry refer to structural evolutions of the competitive forces and environment faced by suppliers, partners, competitors and customers in a given industry, allowing incumbents or new entrants to innovate. It can relate to major evolutions of the industry's barriers to entry or of the bargaining power and rivalry of existing players, in some cases driven by the industrial policy of a major country.

Changes in industry are in particular a key trigger of "creative destruction" innovation processes. Innovations triggered by industry changes include the apparition of innovative new entrants, for example, new airlines, new online players in banking or new "national champions" from emerging countries in the consumer electronics, automotive and steel sectors, such as Huawei, Alibaba or Tata.

New Environmental Constraints

New environmental constraints refer to new "rules of the game" for organizations. This type of innovation trigger relates to major trends which make existing institutions and processes inadequate or obsolete and allow new "ways of doing things" to emerge. They include "implicit" rules related to cultural values and political environments as well as explicit rules such as new regulations.

This means in particular that new regulations, often seen as a burden for innovation, can also be a significant trigger: "necessity as the mother of invention".

Innovations triggered by new environmental constraints include car-sharing businesses driven by urbanization and new manufacturing processes driven by environmental regulations, for example, related to the use of lead or CFCs.

So What?

The implication for managers is that they cannot afford to just wait for their R&D and marketing departments to tell them what science can do or what customers want. Identifying innovation opportunities better and faster than competitors means proactively scanning, mining, screening and combining multiple internal and external potential triggers of innovations.

It therefore means systematically questioning, observing, experimenting and networking across the organization and its environment. Finally, it means ensuring that enough resources and management attention is devoted to these tasks, even when in the short term they have intangible benefits but very tangible costs.

4.1.3 From Generating Ideas to Identifying Sizeable Innovation Opportunities

One of the most common myths regarding successful business innovations is that they were born out of the blue, that somebody (a university drop-out) somewhere (in a garage) suddenly had a clear vision of what could be done and then just had to raise money and make it happen.

As a consequence, many firms engage in "fishing expeditions" for innovation ideas, hoping to uncover the hidden business opportunities of the future among their researchers, employees, customers or even the whole population ("the crowd").

Inventions or discoveries can sometimes be related to single creative ideas, made up in somebody's mind. But innovations always relate to the implementation in specific contexts of new combinations. And this takes time, skill and effort. Great rough diamonds might sometimes be discovered, but they still need to be transported, cut, shaped, set in a jewel, packaged, priced, sold and delivered. Innovation is much more about brewing or gardening than fishing or hunting. It takes time, skill and energy.

Even in the case of the legendary "Post-it", it took more than five years between the accidental discovery by Spencer Silver of a weak but reliable adhesive and the design by Art Fry of the iconic yellow sticky note. It took another few years for the resulting product range to generate a significant share of 3M's revenue. Similarly, even "new economy" champions such as Amazon or Airbnb took many years to emerge and had to go through multiple metamorphoses ("pivots") before becoming sizeable business successes.

From R&D to New Business

In the case of R&D-based product innovations, managers and policymakers often underestimate the long journey that lies between inventing a new technology and generating significant wealth, between increasing public spending in R&D and reducing unemployment.

Even when researchers avoid dead-ends and find positive results, the results have to fall within the priorities and scope of their public or private sponsors. Then tangible knowledge has to be created from the findings and appropriated by organizations. The organizations must be able to integrate the knowledge into viable business opportunities and the businesses need to have sufficient ambition to develop and scale up. Finally, the successful business (if any) has to fit into the corporate portfolio or home region. The complexity of this technology transfer value chain explains why turning R&D ideas into value creation opportunities takes time, skill and effort (and luck).

Replicating the infamous Xerox Park launched 20 years earlier, "Interval Research" was launched by Paul Allen, co-founder of Microsoft Corp., and David Liddle, a computer industry veteran with deep roots in research. After millions of dollars spent on developing cutting-edge technologies, the initiative was shut down, because converting the research results into business opportunities proved (again) much more challenging than expected.

Having Babies or Adopting Teenagers

Faced with the skills, effort and time needed to convert innovation ideas into potential business opportunities, organizations must consider whether to engage in this process ("have a new baby") and manage the implications, or to shortcut it by looking directly for more mature innovation opportunities ("adopt a teenager"). Both options are potentially valid but have vastly different implications in terms of cost, risks, skills and time horizon.

Too often executives expect to generate sizeable revenue out of rough ideas within three to five years. But even the best ideas in the supposedly "asset free" New Economy took five to ten years to become sizable (not to mention profitable) businesses.

In the first case ("have a new baby"), the focus should be on the "newness" side of the innovation coin, on R&D and ideation. It involves investing in order to generate inventions and discoveries with great potential. But it also involves patience and great risks, as the potential opportunities need to be combined and matured, which always takes time.

In the second case ("adopt a teenager"), the focus will be on "change", acquisitions and corporate venturing. It involves investing to identify, acquire or partner with emerging ventures with proven or potential business models, and creating platforms to integrate them and scale them up. This second approach, focusing on acquisition and time to market rather than ideation, is one of the hidden secrets of successful innovation management by large corporations.

Most of the innovation "successes" related to Google outside its core advertising business, in smart home applications, self-driving cars, artificial intelligence, navigation and so on were actually started by acquiring existing businesses (such as Android, Nest or Waze) rather than relying only on leveraging the creative ideas of its own employees and their "20%" free time.

So What?

The journey from new ideas to sizeable innovation opportunities is long and fraught with risks, hurdles and difficulties. Too often executives and policy-makers dream of magic ideas turning into profit and jobs in only a few years. As a consequence, managers must carefully balance ideation initiatives with more short-term/high-impact partnerships and acquisitions. It is in most cases easier to grow one business from $1 million to $10 million than to create ten one-million-dollar businesses.

4.2 Foster Organizational Learning: Beyond Ideation

Potential innovation opportunities can be uncovered when organizations learn new things, when they collectively search new ways to increase their collective and individual "intellectual capital" (Fig. 4.3). This means both having the freedom and finding ways to challenge existing mental models.

In particular, employee expertise and scientific technologies are two corporate knowledge assets that should be managed and developed to foster organizational learning, provided they are harvested in a focused and systematic way and in line with the organization's strategic challenges and priorities.

Key Insights

i. *Organizational learning* is about growing the intellectual capital of the firm across its people and teams, mobilizing problem-driven, opportunistic and systematic search behaviors.

ii. To uncover potential innovation opportunities, organizations and employees need to learn how to generate ideas by *thinking in new boxes*, challenging their prevailing mental models and "reinventing new wheels".

iii. *Employees* can be a rich source of learning and innovation provided that goals and expectations as well as coaching, selection, follow-up and feedback processes are carefully managed. Even in the best organizations most ideas end up being rejected. What matters most is what actually does happen the day *after* the "ideation" events.

iv. Organizations must continuously develop their technology base—manage R&D operations—but also know when to build *new technology platforms*—deploy an R&D strategy—as formerly irrelevant or emerging knowledge and skills become core or even distinctive.

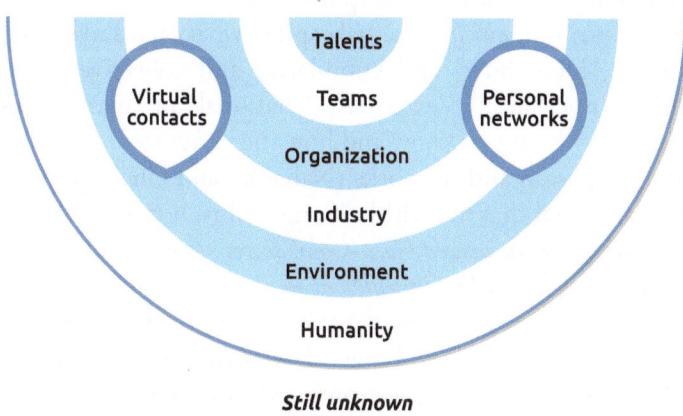

Fig. 4.3 Learning across organizations

4.2.1 Organizational Learning: Growing Intellectual Capital

A probably apocryphal story: an executive congratulated a retiring worker and thanked him for having helped the firm with both hands over so many years, to which the worker supposedly replied, "Too bad, for the same price you could have used my brain too".

Today, firms cannot afford to rely solely on a strong pair of hands. Their performance is increasingly related to their intellectual capital: what they know and how much they learn.

Corporate downsizing and restructuring initiatives often mistakenly lead to invisible losses of critical knowledge, as (pre-)retired workers, frustrated experts or sacked middle-managers leave with what and whom they know. They are indeed sometimes quickly (and expensively) rehired as external consultants.

Know What You Know

Most organizations actually know much more than what they think they know. Of course there is "structural" intellectual capital: what sits in corporate database, software and documents. This includes items such as procedures and quality processes, manuals, trade secrets, patents and publications. In other words, the "structural" intellectual capital represents what somebody could find in a firm even if all the employees left.

But there is also "human" intellectual capital: what each and every employee knows, his/her "know-how" and cumulative and collective competencies. Finally, there is the "social", "external", "relational" or "market" intellectual capital, related to whom you know and how they know you, including brand, reputation, recognition and networks (Nahapiet and Ghoshal 1998). This is the intellectual capital that lies within living brains, not stored documents.

Even if a billionaire uncovered Coca-Cola's secret recipe or Google's search algorithms, the knowledge would be insufficient to replicate their corporate success. As another example, in many professional services and industries, "who you know" is often as important as "what you know". Similarly, even a scientific researcher should be hired based on his or her network, not only his or her expertise.

Revisiting Trials and Errors

The most common and probably oldest way to learn is through experimentation. People and organizations set implicit or explicit intentions and goals, engage in activities aligned with them and then determine whether the results are acceptable. When they are not, they often simply alter their activities, change the facts and hope for better results. This simple exploitation process is often referred as "single loop" learning.

Leading online firms such as Amazon and Facebook constantly tweak their websites and test user reactions to multiple adjustments (so-called A/B testing). Similarly, consumer goods and media firms constantly test the reactions of consumer to specific product offers or adjustments.

In some cases, organizations will go beyond adjusting the initial course of action and also question the original intentions and goals. Rather than "doing the same things better", they consider altering their initial hypothesis, beliefs and mental models. They explore and maybe "do new things".

This latter "double-loop" type of learning can lead to the identification of more radical innovation opportunities, while the former is more related to incremental innovations.

Most people tried to improve the comfort and efficiency of horse carriages by breeding better horses. Then Henry Ford famously thought about replacing horses with an engine. As another example, Sony tried for years to reduce the bulkiness of its TV sets by designing smaller and smaller cathode-ray screens; Samsung instead invested in new technologies (LCD, LED, OLED, QLED) to design flat-screen TVs.

Search Behaviors (Greve 2003)

On top of the "problemistic" approach described above (someone searches for a new solution because there is a problem), people and organizations can also engage in systematic ("institutional") and opportunistic ("slack") search behaviors. "Institutional" search behavior is carried out by dedicated organizational units focused on exploration, such as corporate R&D departments. They learn not only because they have to solve given problems but also because they expect future benefits from the knowledge they will generate. "Slack" search behavior happens when unused resources (such as staff overtime or implicit knowledge) are invested in new solutions to improve performance; learning occurs because time and resources are available for some people to experiment and improve.

Google is famous both for its "moonshot" projects (institutional search) and for the time its staff can dedicate to pet projects (slack search), even though the business impact of both is uncertain. On the other hand, the X-prize competition fosters innovation by challenging people to "push the limits of what's possible to change the world for the better," by finding new ways to solve tough problems (problemistic search).

So What?

Managers should nurture and grow their intellectual capital as carefully as they manage their physical assets or cash. This means gaining a better understanding of "what we know" as an organization, from a structural but also human and social point of view. It also means sometimes challenging not only results and activities (learn to do the same things better) but also initial intentions and beliefs (learn to do new things). Finally, it means combining multiple search behaviors, not only problem driven but also resource (slack) and exploration driven (institutional).

4.2.2 Idea Generation: Thinking in New Boxes

When a problem is identified and/or when available resources make learning possible, old ideas can resurface or new ideas can be generated. There is no scientific consensus regarding the best technique for optimizing this creative process: traditional brainstorming approaches are actually known to be often quite ineffective.

However, it is clear that this creative process can be stimulated and facilitated. Idea generation works better when people have intrinsic (rather than extrinsic) motivation, find meaning in what they do and can mobilize their expertise, and when positive and negative emotions are funneled.

From pure brainstorming to wearing hats, from TRIZ to Lego, from metaphors to adverse thinking, from individual meditation to inhaling exotic substances, multiple techniques have been invented, promoted and used to stimulate creativity and idea generation. What works best probably depends on the people, context, type of problem and above all the skills and experience of the process facilitators.

Whatever the technique, fostering the process requires getting people to move outside their mental comfort zone—thinking inside new boxes—and helping them design new potential answers—thinking about new things.

Thinking Inside New Boxes

Engineers are known to often be unable to think "out of the box" until the specifications (size, shape, thickness, dimensions, etc.) of the box have been clearly defined. This anecdote shows that it is actually impossible to think outside any boxes. Thinking requires mobilizing words, images, interpretations, cultural values, knowledge. What is thought about is framed by these "fixations". Thinking requires mental models, which allows people to frame the world and apprehend it (Kaplan 2008). Thinking requires boxes.

Individuals therefore construct new knowledge from their experiences by assimilation, incorporating the new experience into an already existing framework or mental model—the existing box. But they also construct new knowledge by accommodating, bending, transforming and reorganizing the cognitive constructs that represent their view of the world and reframing their mental representation—the new box.

Therefore, what can be done to foster idea generation is changing the mental models, the perspective, that is, thinking in new boxes. In this sense, creativity is not a process of "generating" new things out of the blue but rather "seeing what is already there" and that our existing mental models, acting as "blinders", prevented us from seeing. The challenge is being aware of existing "blinders" and being able to remove them. Creativity is about looking at existing things in new ways.

Gravity and relativity existed before Isaac Newton and Albert Einstein "discovered" them. It took 130 years for Heinz to put its famous ketchup bottle upside down (to minimize waste), but the possibility had always been there. Apollo XIII had only a few minutes to invent a new filtration system and save their lives, but all the elements were of course already on board.

Thinking Inside New Strategy Boxes

From a corporate point of view, an important driver of opportunity identification will therefore be the "strategy box". This relates to what members of the organizations implicitly or explicitly see as the "legitimate paradigm", the "natural" activity of the firm, its "core business". When identifying innovation opportunities most people will (consciously or not) "stay in the box" in order not to come up with irrelevant or "silly" ideas. What they see as the limits of that box and whether there is a consensus regarding these limits therefore clearly matters. Defining this opportunity landscape too widely will lead to frustration and wasted time as irrelevant avenues are explored. But defining it too narrowly will lead to conservatism and missed opportunities as potential synergies are discarded.

In order to complete a complex jigsaw puzzle, the best approach is often to start with the border.

Depending on the firm, most managers would agree on the things it will never produce, like pralines or spaceships. What they actually see as its core business ends somewhere between what it does today and clearly out-of-scope ideas like pralines or spaceship. The key question is therefore: "Where exactly does the core business end?"

Nokia famously completely redefined its core business several times, from raw materials to telecommunication equipment. As another example, some car companies see their business now more as "providers of mobility" (including renting bikes, software platforms or public transport) rather than "car manufacturers".

Reinventing New Wheels

Once a "new" box of innovation opportunities has been defined and opened, the next challenge is to identify many attractive opportunities within that new box. A powerful way to do so is by "reinventing new wheels", that is, mobilizing extensive existing knowledge to generate new ideas. As innovations are always new combinations, the greater the number of existing ideas that can be mobilized, the greater the number of new combinations.

Contrary to popular belief, children are often much less creative than adults, because they know much less (most children will draw people, cars, houses or trees in very similar ways). However, they have much lighter "frames" and therefore are not afraid to speak their mind and come up with "crazy" ideas that adults would not dare suggest.

New combinations can be generated in a systematic way by first turning a concrete situation into an abstract functional problem ("emptying that glass" becomes "moving a liquid"), then scanning and screening the myriad of existing solutions to that abstract problem (evaporation, shock waves, etc.), then finally selecting the existing solutions (the "wheels") that could be transposed to solve the initial problem.

Organizations such as IDEO and CREAX have developed systematic approaches to "invent new wheels", by scanning patents, scientific publications and the deep web to uncover useful sources of inspiration; CREAX's slogan is "Somewhere, someone has already solved your problem". Another example of "inventing new wheels" is biomimicry, that is, exploring ecosystem biology to see how "nature" has solved specific problems and use its solutions as inspiration. Examples of biomimicry include Velcro (inspired by burrs), swarm intelligence (inspired by ants and bees) and dust-resistant paint (inspired by lotus leaves).

So What?

Idea generation is not a mysterious process that requires black magic-type approaches. The identification of innovation opportunities by people and teams can be fostered when on the one hand they can change their mental models and explicitly "think in new boxes" and on the other hand when they can effectively leverage existing knowledge in order to "invent new wheels". To do so, managers should focus on picking the technique that fits their culture and priorities, creating a meaningful and motivating working environment and mobilizing the right facilitators.

4.2.3 Fostering Employee-Driven Innovation

While marketing and R&D departments are often seen as the "natural" sources of the new ideas emerging from customers ("pull") or technologies ("push"), a firm's intellectual capital actually resides throughout its organization. Learning and innovation opportunities can be found across all functions and departments. From the old idea boxes to sophisticated corporate social networks, fostering such employee-driven innovations (EDIs) has therefore the potential to be a valuable source of innovation opportunities.

EDI approaches include passive idea collection processes ("idea boxes") and more targeted campaigns based on specific organizational, market or other challenges. They can rely on punctual calls to employees or on more active communication tools such as communities of interest, discussion groups, blogs, and so on.

Employee-Driven Frustration

One of the issues with EDI approaches is that their popularity is probably proportional to their lack of effectiveness. Too often organizations go "fishing for ideas", hoping that the next business success will spontaneously emerge from somebody's drawer. But good ideas almost never emerge spontaneously from individuals. They need time to sprout and mature. They are nurtured by interactions and confrontations.

Furthermore, presenting a good idea in a convincing way ("the pitch") is a difficult task that is beyond the skills of most untrained people. Finally, even when well presented, only a fraction of ideas will be selected and even fewer will be implemented, leaving many "inventors" potentially frustrated.

Among all the ideas pursued by Google employees during their well-known "20% time", less than 1% become actual projects and only a handful have led to significant business impact. Innovation is 1% inspiration, 99% perspiration.

Start with Why

One of the key challenges of implementing effective EDI approaches is the management of expectations, in terms of both objectives and playing field. In terms of objectives, EDI can be used either to foster employee engagement or to identify high-potential business opportunities. Both expectations are valid, but they involve completely different approaches. In the first case, the focus will be on participation rates, interactions and feedback. In the second case, the focus should be on "picking winners" and aggressively scaling them up.

Overall, these initiatives tend to have limited direct effect on corporate economics but can create healthy emulation and strong positive effects on internal "atmosphere", networking and external reputation.

EDI approaches focused on employee engagement are like amateur Olympics, where "the most important thing is not to triumph but to compete!" EDI approaches focused on high-potential business opportunities are more like beauty contests, when only the winner gets remembered. Both approaches can be valid but trying to combine them is probably the worst choice.

In terms of "playing field", EDI will be more effective when the "new strategy box" inside which people should think is clearly outlined, be it in terms of perimeter (what kind of ideas), time horizon, levels of ambition or key metrics. Managers should not expect employees to guess what kind of ideas they are looking for. Ambiguity is not a substitute for flexibility. Creativity is fostered when the work done has meaning and when the relevant skills can be mobilized.

From decreasing carbon footprint to increasing customer satisfaction, EDI can be used for a wide range of objectives. But not clarifying the objectives upfront often leads to frustration and inefficiency.

Finally, the process should be facilitated by communities of practices, tools, training and people, allowing people with potentially good ideas to understand how and where to improve and submit them. Presenting an idea in a meaningful and convincing way is not innate.

Focus on the Day After

The second key challenge related to EDI is to remember that innovation is much more than ideation. The most important part of an EDI approach is therefore not the harvesting of ideas but what happens next. In particular, are the people who submitted the winning ideas supposed and/or allowed to pursue them, and with what mandate and resources?

Too often the focus of EDI initiatives is on big "innovation events" (hackathons, business plan competition, jams, acceleration programs, etc.) after which most people "go back to work". As Harvard's Theodore Levitt already noted in 1963, "Advocates of creativity must understand the pressing facts of the executive's life: Every time an idea is submitted to him, it creates more problems for him – and he already has enough".

Finally, even with the most creative employees, the large majority (typically more than 90%) of ideas will be rejected as unfeasible and/or irrelevant. That leaves many potentially disappointed and/or frustrated people if not enough attention and resources are devoted to providing them with timely and actionable feedback.

The worst cases of EDI occur when both good and "bad" ideas are not adequately followed up, resulting in both no positive business impact and decreased employee engagement. Grassroots are great but must be gardened.

So What?

"Employees are our best asset" is a valid but overused motto. EDI approaches can be effective tools, in particular to stimulate employee engagement, provided the right approach is put in place in terms of setting the scene, gathering and assessing ideas, ensuring adequate follow-up of both accepted and rejected ideas, and monitoring and adjusting the process along the way.

Ideas are cheap. What matters most is what actually happens the day after the ideation events have been organized.

4.2.4 Developing New Technology Platforms

Most firms have an explicit or implicit list of competencies that they consider key to the competitiveness of their business. They therefore tend to invest significant time and energy in benchmarking and developing "technology platforms" and keeping their skills up-to-date.

But beyond the maintenance of their existing technology platforms, firms must also decide which new technology areas they want to engage in and which technology areas they want to freeze or completely abandon.

R&D Strategy: Deciding What Not to Do

Businesses must ensure both the maintenance of core technology competencies and the continuous development of the distinctive technologies that currently allow them to differentiate themselves from their competitors.

But being good at what matters today is not enough. On top of the management of "R&D operations", managers must also decide on an "R&D strategy". This means deciding which future technology (r)evolutions they want to lead, which they want to just follow and monitor, and which they want to leave aside for the moment. This means understanding which of the existing or emerging technology platforms will become core or distinctive elements of their competitive roadmap, and which will remain or become irrelevant.

In the car industry, existing technologies such as electric batteries or new technologies such as sensors and connectivity are becoming key differentiating factors. Automatic gearboxes and diesel engines have been critical technologies for many years but might become irrelevant as cars become hybrid or fully electric. The software part of a vehicle used to be there just to control key hardware but is expected to create most of a car's value in the future. Finally, hydrogen-based and fuel cell technologies have been "around the corner" for many years but might (or might not) remain so for another decade.

R&D Challenges: Executing the Strategy

Having designed an R&D strategy, managers must design and develop the R&D organization that will implement that strategy. But the challenges of R&D organizations now go far beyond effectively managing the key experts, projects and know-how. On top of choosing in which technology platforms and competencies to invest ("what"), they must design the right governance structure ("where") and monitor the level of investment ("how much").

In terms of governance, the first trade-off is to decide how much technology development will be directly controlled by the firm and its partners and how much the firm will rely on the orchestration and coordination of R&D activities performed by others ("make or buy"). R&D partners can be its suppliers of equipment, specialized technologies and niche products or providers of data and key information.

Telecommunication operators and large aircraft producers today rely much more on the integration of equipment and software developed by their networks of suppliers ("original equipment manufacturers") than on their own "R&D" activities, if they have any.

The second trade-off is the level of centralization of technology development. Centralized hubs of corporate knowledge can create synergies and reach critical masses of expertise by focusing on specific technology domains and/or application markets. Decentralized structures allow organizations to better adjust to local operating conditions and business needs, such as regulations or raw material availability and cost.

Being present in many places can also allow an organization to tap multiple local sources of knowledge, such as local customers, technology clusters or informal expert networks.

While less than 20% of countries still account for more than 80% of global R&D spending, most technology-oriented firms must now find ways to have access to the knowledge developed across key areas such as North America, Europe or Southeast Asia, and increasingly China and India.

The last but not least trade-off relates to the balance of investments between the different technology platforms, R&D structures and geographies. Too much investment and the quality of delivered R&D projects will not justify their costs; too little and the option value of business opportunities opened by technology development will be lost. While the "right" level of R&D spending seems to vary significantly across sectors and is constantly benchmarked by many firms, just spending more across R&D is definitely not an automatic recipe for success.

Innovation managers must remain humble regarding their R&D investments, and remember what Einstein supposedly said: "If we knew what we were doing, it would not be called research".

So What?

On top of ensuring that they effectively run their existing R&D operations, managers must identify the technology platforms they want to lead, follow or leave aside, in other words manage their R&D strategy. From an organizational point of view they must also balance what they develop centrally versus locally and through third parties, combining a critical mass of expertise with local and external sources of knowledge. Corporate R&D centers still matter but most cannot be effective on their own.

4.3   Harvest and Protect Organizational Knowledge Assets

Tapping various sources of innovations and fostering learning, technology development and ideation can allow firms to create new knowledge and develop their intellectual capital.

But innovation is based on new combinations, not only on new knowledge. And competitive business is based on differentiation, not just doing and knowing what everybody else has already done and known. Effectively identifying innovation opportunities therefore requires organizations to find ways to systematically harvest and protect their knowledge assets (Du Plessis 2007).

This means on the one hand understanding the peculiarities of knowledge as an asset to be developed and absorbed, and on the other hand designing and executing the right strategy to protect and maintain this knowledge as a source of competitive advantage (Fig. 4.4).

Key Insights

  i. To identify attractive opportunities, organizations must systematically manage their explicit and implicit *knowledge assets*, finding ways to share, combine, disseminate and maintain this unique type of capital.

  ii. Innovation opportunities can be sources of competitive differentiation provided that the intellectual capital they are based on is actively *protected* through a mix of secrecy, lock-in and/or intellectual property rights.

 iii. Innovation opportunities can be leveraged if they are supported by effective *patenting* strategies, balancing protection and disclosure. Not everything that can be protected should be, and patents can and should foster rather than hinder collaboration.

Fig. 4.4 Managing knowledge as an organizational asset

4.3.1 Managing Corporate Knowledge Assets

Knowledge is not just created by people and organizations. It is also captured, shared, combined and transformed, and this process is at the core of the emergence of innovation opportunities. Knowledge management therefore entails systematically identifying, acquiring and developing new knowledge, and creating "knowledge intensive" firms; the focus is on managing people and what the firm itself has invented or developed. Knowledge management also entails proactively using, storing and sharing existing knowledge; the focus is then on managing processes and what others have invented or developed. A firm's intellectual capital is therefore not driven only by "what it knows" but also by the tools and processes it has put in place for knowledge and expertise to grow (Grant 1996).

Internet firms often identify, acquire and develop knowledge about new online processes and the way they handle users and manage products and services. Traditional retailers, on the other hand, must use, store and share knowledge regarding how these new processes can be implemented across their own organizations. Similarly, most incumbent financial institutions had to manage innovations that were introduced by others, such as Amazon, Facebook or eBay.

But most knowledge cannot be manufactured, distributed and stored like common physical assets. Organizations must therefore make sure that they put in place environments and approaches that help them learn together, that is, to effectively share and combine knowledge among their members. They also need to learn quickly and leverage existing ideas. Finally, they must learn to learn, that is, learn to minimize wasted or lost knowledge.

Professional firms and retailers are examples of organizations that invest a lot in developing and exploiting new knowledge, be it contextual information or expert insights, embedded not only in procedures and documents but also in routines, processes, practices and norms.

Knowledge Is Pow(d)er

Knowledge is a key but peculiar type of asset for a firm to manage. First, it is often only available as "tacit knowledge", that is, knowledge that is subjective, experience-driven, partially unconsciously acquired and maintained and therefore difficult to codify and formalize. It is therefore often difficult to transfer knowledge from one part of the organization to another.

Second, knowledge is often idiosyncratic, as it is aggregated, developed and adjusted to specific contexts. What is relevant or key in one situation might not be in another. It is therefore sometimes difficult to generalize and share corporate knowledge.

Third, knowledge is by nature a "public good", in the sense that consumption of it by one person does not reduce the amount available to be consumed by others. It is therefore difficult to appropriate knowledge and to use it to create competitive advantages.

Finally, more knowledge is not always better. There is a limit to how much knowledge people and organizations can handle effectively, and specialization is therefore a challenge.

Many start-ups based on the scientific know-how of laboratories fail to succeed and scale up. This is not only because they lack marketing skills or funding but also because they cannot find effective ways to diffuse and monetize the knowledge embedded in their experts.

Critical Knowledge: Less Is More

Effectively managing knowledge means much more than accumulating information in corporate archiving systems. It means understanding and locating the knowledge and data that is critical for the organization. It can be critical because its scarcity and relevance make it a vital element to its competitiveness, or because gathering, codifying, diffusing, using, updating and leveraging it is a significant competitive barrier to entry.

Organizations must therefore identify the critical knowledge domains on which to capitalize and the key people and expertise to retain as a consequence. They must also identify the best ways to share and exploit but also preserve that specific knowledge.

Industrial organizations often discover that some vital know-how is possessed by a limited number of sometimes undervalued experts. They must therefore implement procedures and systems to systematically identify and proactively manage these experts, as well as find ways to acquire, develop, use, distribute and preserve their knowledge.

Absorbing and Mediating Critical Knowledge

Anyone with an email or a social network account knows that being bombarded with information is not the same as acquiring knowledge. In order to be turned into exploitable knowledge, this information must be mediated and absorbed (Nonaka 1994).

Mediation can be facilitated by automatic search and new "intelligent" filtering tools, but it should also be fostered by "gatekeepers": people explicitly or implicitly identified as the organization's repositories of key knowledge. They either possess the knowledge or know "who knows" and can create and facilitate useful connections, acting as a go-between organizations or parts of the same one. Facts and figures matter, but so do remembering faces and referring people.

The absorption and development of the explicit and implicit knowledge available in an organization must also be systematically fostered. Explicit knowledge can be combined through analysis and computations, in order to create new explicit knowledge. It can also be internalized through learning and experimentation, in order to create new implicit knowledge. On the other hand, implicit or tacit knowledge can be "externalized" through reporting, training or the production of prototypes and mock-ups, in order to create new explicit knowledge ("people-to-document"). Finally, implicit or tacit knowledge can also be shared through "socialization" activities such as networking, teamwork or apprenticeship, in order to create new implicit knowledge ("learning by doing", "people-to-people"). While the direct impact of all these activities is often intangible and difficult to measure, they matter as much or sometimes even more than the direct production of goods or services.

Sharing experience across business entities is a way to mediate and absorb both explicit and implicit knowledge. "Big data" and "artificial intelligence" applications are examples of new knowledge combinations, creating new explicit knowledge from existing ones. The challenge for many organizations is to be able to blend explicit knowledge tools with their existing (mostly tacit) knowledge.

So What?

The knowledge of an organization is in itself a key but peculiar asset, which must be developed carefully. This involves first implementing the right information management systems and procedures. More important, it also entails developing a work environment, network of "connectors" and corporate culture that foster not only knowledge codification but knowledge creation, formalization, protection, sharing and absorption. Knowledge management is a people issue with technical aspects, not the other way around. Finally, it means recognizing that the management of critical knowledge and data, that is, identifying both the knowledge and how to deal with it, matters as much and sometimes much more than the production of the goods and services it supports.

If you think that learning is too costly and time-consuming, try ignorance (adapted from Char Meyers).

4.3.2 Protecting Intellectual Capital

Knowledge is like fresh air: in most cases consuming it does not reduce the amount available to others. While this has excellent social benefits, it means that an individual firm might have limited incentives to develop new knowledge beyond purely altruistic motives. If significant investments are needed to turn some intellectual capital into a successful innovation, specific protection mechanisms might therefore be needed in order to turn a "public" good into a "private" one (Samuelson 1954). This is particularly the case for explicit (codified) knowledge, which is often easier for a competitor to capture and use.

Some known medical treatments are not made commercially available because a firm that is investing significantly in their validation, sales and marketing would not be able to secure the exclusive rights to profit from the results. However, significant innovations such as "Wikipedia" have been able to rely only on social norms and altruistic behaviors to support their development.

Fencing: Intellectual Property Rights (Pisano 2006)

Organizations that invest in land, facilities or equipment can in most organized economies secure the rights to use the land, buildings or machines. But the knowledge they create cannot easily be "fenced" like a patch of land or tagged like a machine. Most developed countries have therefore developed specific regulations protecting the ownership of specific intellectual assets. These regulations aim in most cases at providing a temporary private benefit to an inventor, in the long-term public interest.

Food companies that invest in medical studies that prove a specific ingredient has significant health benefits can in some cases obtain an exclusive protective right to use references to the results of those studies in their marketing. Brands and URLs are other examples of knowledge assets that can be protected.

Regulations can aim at protecting investments in creativity, such as artistic creation, designs or know-how, or in reputation, such as trademarks. They can be provided automatically and for free, for example, in the case of copyright, or require specific registration steps and payments, for example, in the case of registered designs. Some "creative" industries such as media, software, design or publishing have specialized in the generation and exploitation of such intellectual property, based on individual creativity, skill and talent.

Disney and more recently Nintendo have based their success on marketing and exploiting through movies, cartoons, entertainment and merchandizing their portfolio of famous characters. Brands such as Coca-Cola, Google, Apple and Amazon are worth billions of dollars and are therefore actively protected. Intel is known to aggressively defend its "Intel Inside" slogan, even in industries not related to its core business.

The Wild Wild East

Intellectual property rights are regulations and as such must rely on effective institutions to enforce them. Intellectual property rights are therefore by nature attached to specific geographies over which institutions have jurisdiction. The geographic scope of rights must be assessed and chosen carefully, as local authorities can be unreliable, corrupt or biased in favor of local champions.

Many emerging countries and some US states are known to be challenging places for foreign firms to have their property rights effectively enforced.

Existing property rights can also be used by incumbents to fight and hinder the development of new firms. Even if the underlying claims might in some cases be ultimately proven baseless, the resulting legal processes and reputational effects can be sufficient to kill many new ventures, given the cash and time needed to cope with the legal proceedings. In some industries the cost of just securing such "freedom to operate" can itself be a barrier to entry.

This implies that intellectual property rights can also be an obstacle to innovation, slowing exploration and adoption or leading to the duplication of research activities. The actual net economic benefits of intellectual property regulations have been hotly debated, and the existence of such rights has been regularly challenged since their inception in the eighteenth century.

Certain technology areas such as gene editing or mobile telephony are considered "intellectual property minefields", where new players are quickly bombarded with costly lawsuits. In such industries, deep pockets and strong bargaining power are needed to innovate. As another example, the once dominant De Beers diamond company has been aggressively fighting, through property right litigations, new entrants that are developing artificial diamonds.

Great Honors Are Great Burdens

The challenge and complexity of obtaining and enforcing intellectual property rights has led many organizations to develop alternative strategies for protecting their knowledge, either by hiding it or by protecting it through other means than property rights.

Coca-Cola's recipe, the Firefox browser code or the "Post-It" glue formula are all examples of intellectual capital assets protected by alternatives to property rights. According to the European Commission, only about 10% of major industrial innovations are patented, suggesting that the great majority rely on secrecy or other types of competitive advantages.

The most common alternative to intellectual property rights is the maintenance of trade secrets. While this has potentially perpetual benefits, it requires a very systematic and tightly controlled management of sourcing, documents and facilities, staff turnover, training and poaching as well as preemptive protection against competitors claiming intellectual property rights on the firm's secrets.

New surveillance technologies, hacking by sophisticated players as well as staff rotation and extended supply chains are examples of the formidable challenges faced by firms that rely on secrecy to protect their intellectual property.

Attaching Strings

Another way to protect a firm's intellectual capital is to tie its value to other activities or assets. In this case, a competitor might access the underlying knowledge but will struggle to replicate the activities or assets. One way is to embed a firm's intellectual capital in costly early investments in manufacturing capacity, making competitive replication attempts obsolete by the time they are ready. An alternative is to develop unique complementary manufacturing, sales or services that would be difficult or impossible to replicate.

The huge investments of Intel and Tesla in respectively chips and battery manufacturing are difficult for competitors to replicate before the corresponding technologies become obsolete. As another example, so-called open-source business models actually actively share their knowledge but achieve competitive differentiation through complementary services such as sales and maintenance or integration.

So What?

Innovation managers must protect their intellectual capital if they do not want competitors to threaten or free-ride their inventiveness and creativity. This means on one hand maintaining freedom to operate and on the other hand managing the right combination of intellectual property rights, secrecy and development of complementary assets and activities.

4.3.3 Managing Patents

One of the best-known and most common ways to protect knowledge related to a technology via intellectual property rights is the patent, where an inventor exchanges disclosure against a temporary local monopoly. But patents are often misunderstood tools and should not be considered a universal panacea.

Patents

A patent usually protects a technical creation that fulfills three conditions: novelty, or not having been publicly disclosed before; inventive step, or being non-obvious; and usefulness, or applicability at industrial level. It is granted by a government and is limited in time and scope, both from a market and geographical point of view. Patents provide incentives to innovate but can also be used to limit the "freedom to operate" of new entrants.

Patents can therefore be very useful. But they are also often misunderstood, in terms of the rights they provide, the validation they involve and how often they should be used.

First, a patent is not a right to make or sell anything but only a right to exclude others. That is, a patent allows a firm to prevent (via litigation) others from making, using, marketing or selling the disclosed invention for a certain period of time, usually 20 years, but it does not bestow on that firm alone the right to make, use, market or sell that invention.

A firm might be able to patent the use of a famous smartphone as a toothbrush, but that would not allow it to make and sell those toothbrush-enabled smartphones to consumers. A patent is a right to sue, not a right to use.

Second, a patent does not provide a validation of the business potential of an invention. It is not based on an assessment of its quality or performance but on its perceived novelty vis-à-vis the state of the art. The quality and number of patents filed for the same type of invention can actually vary wildly from one industry to another and from one country to another. Patents evaluate inventions, not innovations.

There are multiple patents on love and perpetual movement, which as far as we know is impossible. There is even a patent for a lightbulb changer and a patent on the use of patents. Yet "number of patents" is still used erroneously to measure innovativeness.

Third, not everything that is patentable should be patented. Patents have benefits and costs related to the underlying technology, competitive environment and legal context, all of which need to be assessed. Some innovations are and should be protected by patents, but some should not.

The shape and color of a "Post-It" note are protected by intellectual property rights, but its glue has not been patented and remains a trade secret.

Benefits of Patents

The basic trade-off involved in a patent is disclosure against temporary and local monopoly. The main benefit of a patent is therefore the economic value of controlling the development of a technology and excluding competitors from certain markets for a certain amount of time.

Pharmaceutical companies have for decades relied on their patent-related monopoly on the sales of some blockbuster drugs. Firms like Amazon, Microsoft and Apple have patented hundreds of ICT-related business processes in order to (try to) prevent their competitors from using them.

There are, however, other benefits that should be taken into account. First, a patent turns an intangible knowledge asset into a tangible property right. This means that like any other tangible asset it can now be sold or used as collateral or as a contribution to a partnership agreement. It can also in some cases generate significant tax benefits.

Qualcomm has based its success on the ownership of key patents related to mobile phone technologies, although it does not itself sell mobile phones.

A common way to monetize such "tangibilized" knowledge assets is through licensing, where intellectual property is transferred to a third party under specific exclusivity, scope and reward conditions. While sometimes complex to define and manage, such agreements allow firms to share costs and risks, reach new markets and support specific standards.

IBM has long generated significant revenues from the licensing of technologies it invented but did not want or was not able to market itself.

Second, a patent allows a firm to actually disclose and market what it does. This disclosure can be leveraged as a signal of innovativeness or as a source of bargaining power. It can also be used to reward creativity and boost staff motivation. The disclosure of protected intellectual property can facilitate cooperative research activities, by preempting leaks and preventing unwanted appropriation.

Research collaborations where the ownership and claims of each partner do not overlap, are clear and are stated upfront can be easier to manage than partnerships shrouded in secrecy and jammed with non-disclosure agreements.

Finally, a patent can protect against other patents. It can boost the firm's freedom to operate, by preventing competitors to patent in the first place or by providing the patent owner with a bargaining position for cross-licensing or retaliation.

Google bought Motorola and part of HTC not because it wanted to manufacture mobile phones but mainly because it wanted to boost its bargaining position vis-à-vis the intellectual property rights of other mobile technology players such as Samsung or Apple.

Patent Drawbacks

Patents provide benefits but also have significant drawbacks, which should be carefully considered. The main drawback of a patent is the disclosure it implies. As the patented technology becomes public knowledge, competitors are able to find inspiration, discover new approaches or invent around existing patents. More generally, patents can provide insights and hindsight regarding the strategy and product development priorities of the patent holder.

Most technology-intensive firms closely monitor the patents published by their competitors, and use this monitoring as a valuable source of technology and competitive intelligence. As a countermeasure, firms sometimes promote fake patents or irrelevant technology claims in order to confuse competitors and hide their real technological breakthroughs. They can also "camouflage" their property rights by applying for them in exotic jurisdictions or only in local languages.

The less obvious but not less important drawback of a patent is its reliance on the detection and litigation of infringements. In some cases, infringements can be difficult or impossible to detect, because they relate to new products sold far away or to new processes implemented deep inside competitors' facilities.

Even when detected, litigation can be impossible or very costly, in particular when dealing with weak or complex legal systems. The full cost of a patent is therefore not only the cost of filing and maintaining the patent, which can already represent a significant investment for a new venture, but also the often huge costs related to detecting and suing infringements.

The so-called patent trolls base their business model on aggressively suing firms, leveraging sometimes dubious patent claims, and often hoping that the huge legal and business costs potentially involved will push their target to settle.

From a more general perspective, the proliferation of patents can also sometimes hinder innovation as such, by excessively limiting new entrants' freedom to operate and by slowing adoption by customers and complementors, which can become afraid of locked-in technology.

The development and adoption of gene-editing technologies have been strongly constrained by the conflicts arising among a small number of key patent-holders. Conversely, some "emerging" countries have decided to override existing property rights because they significantly increased the costs of life-saving technologies.

So What?

The management of patents is a strategic, not only operational, decision. Not all patentable technologies should be patented by a firm. Choosing whether to patent in a targeted or comprehensive way, an aggressive or preemptive way, an international or local way, directly or through licensing, involves strategic decisions. Those decisions should take into account the direct and indirect benefits of patents but also their drawbacks in terms of competitive intelligence and costs.

4.4 Integrate External Sources of Knowledge

Whatever the size and expertise of an organization, the world knows more than it does. Organizations will therefore identify better and faster innovation opportunities if they can continuously cross-fertilize their knowledge with what others already know or do, regarding technologies, customers and industries.

This means developing the capability of the organization to connect with new sources of technology intelligence (Fig. 4.5), uncover untapped customer needs and build ties with disruptive new ventures.

Key Insights

i. On top of their own R&D activities, product roadmaps and experience curves, firms must develop the ability to proactively *harvest technology (r)evolutions*. They must be able to absorb technology intelligence from external sources, both in their socioeconomic ecosystem and in the wider environment.

ii. On top of their traditional marketing approaches, firms need to develop user-centric ways to "pull" *untapped customer needs* and uncover value gaps. They need to combine powerful analytics with in situ and empathetic observations as well as lead users' interactions and involvement.

iii. Firms exposed to technology-intensive sectors (i.e. most firms) should invest time, money and resources to network and work with disruptive start-ups, in particular through *corporate venture capital* initiatives. This must allow them not only to create options and leverage their assets but also to effectively develop new competitive and technology intelligence.

4.4.1 Harvest Technology (R)evolutions

As late as in the nineteenth century, some "polymath" scientists could claim to know everything that was then known regarding one subject. But today even the biggest corporate or public R&D lab can only manage a fraction of

Fig. 4.5 Looking out for innovation opportunities

the knowledge developed in its field. Being able to actively connect with, integrate and absorb the knowledge developed outside the walls of its own organization is therefore critical for any organization competing in technology-intensive environments.

This means both identifying where the external sources of technology intelligence can be found and developing the capability to absorb the knowledge they generate.

Sources of Technology Intelligence

R&D activities are still often pictured in a laboratory setting with scientists in lab coats busy with complex equipment. But today developing cutting-edge technology intelligence requires spending significant time outside the laboratory walls, learning what others know and do.

This means tracking online and offline scientific publications and patents, both in terms of their content and relative frequency, and trying to discover emerging "hot" topics. It means buying competitors' products to benchmark and reverse-engineer them and consider licensing-in options. It means interviewing a competitor's employees, sometimes hiring them or acquiring their business. Finally, it means interacting with experts and suppliers of specialized goods, services, technology and equipment in order to discuss potential developments and build roadmaps and scenarios.

Some firms have significantly reduced or even shut down their own R&D activities, relying on absorbing and integrating what others (in particular their suppliers) are developing.

Conversely, firms should systematically assess and manage how much their own activities provide sources of technology intelligence for their competitors.

While most firms are now increasingly aware of the dangers of competitors stealing their knowledge, it is still amazingly easy to gather intelligence by just "walking around" in the right buildings or taking a flight to the right conference. Most "leaks" are still the product not of sophisticated "hackers" but of naïve employees.

In some cases the relevant sources of technology intelligence can even go beyond the firm's socioeconomic ecosystem and reach the wider environment. Indeed, some firms spend significant time analyzing how nature ("greensourcing") or past civilizations have solved specific technical challenges and use those approaches as inspiration.

Examples of "greensourcing" approaches include screening the health and nutrition benefits of the thousands of traditional plants that were once used as feed or food (today fewer than 200 plants cover most of what is fed). Another example is to understand the sources of antimicrobial resistance of exotic species (such as crocodiles) in order to screen them for new antibiotics.

Do Not Outsource Your Incompetence

Learning about a technology from external sources requires in most cases some pre-existing knowledge about that technology, if only to know what you do not know. Firms therefore need to develop the internal capability to ask the right questions and look in the right direction. They also need to be exposed to repeated and effective interactions with others, in order to have the opportunities to pose questions and observe. Finally, they need to learn how to collaborate with individual partners and manage those partnerships, in order to recognize, assimilate and utilize new knowledge to create value.

Such "absorptive capacity" (Cohen and Levinthal 1990) can be supported by a combination of previous knowledge, experience and complementary technologies, prior investments in learning and a strong and explicit readiness and intention to learn and share.

R&D experts in the telecommunication industry spend a significant amount of time in open technical meetings and specialized conferences. Some operators have even completely shut down their own R&D department, relying instead on developing the ability to absorb, combine and market technologies developed by others. "Proudly found elsewhere" can sometimes be more effective than "invented here".

So What?

Investing in R&D often means hiring experts with impressive PhDs and buying expensive top-notch equipment. But investing in R&D must also mean investing time, resources and capabilities to explore external sources of intelligence by reading, traveling and attending meetings to learn and absorb what others know and do—all of which is increasingly more important.

4.4.2 Explore Untapped Customer Needs: User-Centric Innovation

The traditional way to identify new product innovation opportunities in a market is on the one hand to ask or test customers on what they want and on the other hand to benchmark what competitors already offer. But customers do not always know what they (could) value and competitors do not always focus on the right features. Effectively identifying new product innovation opportunities in a market must therefore include reaching out and involving users and customers in order to identify unarticulated needs and untapped value gaps.

Value Gaps

A value gap can appear when customers and/or competitors overemphasize some features, such as speed, weight and shelf life, at the expense of others. In such cases incumbents will continuously out-compete each other in a limited number of specific dimensions, leaving other valuable existing or new features underserved.

Value gaps typically arise in industry with rapid technology evolutions, where competitors overemphasize technology-driven features at the expense of more user-centric dimensions. They can also arise in service industries where quality as defined and measured by the suppliers or their management can significantly differ from users' perception of what is actually delivered.

The automotive and telecommunication industries' continuous focus on ever-increasing technology sophistication and performance can leave value gaps related to untapped customer needs, for example, regarding convenience or sustainability. As another example, utilities focused on overall capex or downtime reduction sometimes fall short of what customers expect in terms of customer service or reliability. Finally, the race of retail-banking actors for more automatization and virtualization and less customer-facing time can open value gaps for more consumer-centric firms to exploit.

Value gaps therefore offer opportunities for new products differentiated along other or new features, sometimes even allowing for lower performance along traditional features. Exploiting them, however, requires successfully identifying the untapped needs and convincing customers that the focus on the new features do not excessively disrupt their routines.

The initial incumbents of the mobile phone industry (Nokia, Blackberry) were disrupted by new entrants, such as Apple and Samsung, that focused on value gaps related to applications platforms and data services. Nokia and Blackberry then tried to reenter the market by exploiting other potential value gaps left by Apple and Samsung, such as limited battery life, low-quality cameras or lack of privacy.

Uncovering Value Gaps in a Market

While there are no magic instructions for uncovering value gaps, there are opportunities for many firms to develop their capabilities to understand the customer problem that they are, or should be, solving. Firms can, for example, better and more often involve customers and end-users in the design and development of their products. Such opportunities include quite old approaches such as quality function deployment and lead users but also new "user-centric" approaches such as "Living Labs", ethnographical observations and artificial intelligence.

Quality function deployment approaches rely on a systematic matching of, on the one hand, the technical parameters of a product (what is feasible from a technical and financial point of view) and, on the other hand, its functional parameters (what can be delivered in a value-added and differentiated way). This matching can allow for identifying new features that are desired by customers and that are at the same time technically and financially feasible.

For example, glass manufacturers had long suspected that windows and glass in general had many untapped features. They also knew that silver had antibacterial properties and might be embedded in glass. They also found out that some of their customers, such as refrigerator manufacturers and hospitals, were suffering from microbial infections of their equipment. The matching of these technical and functional parameters and the resulting value gap led to the identification of new antibacterial glass as an innovation opportunity.

Lead users (Von Hippel 1986) are specific segments of "hyperactive" users, who can in some cases be significantly mobilized in the design and development of a new product. While such involvement of users can create issues in terms of their motivations, the control of the process and property rights, it can also be a valuable and accessible source of insights and innovation opportunities.

Lego is a famous example of a firm that initially resisted (because it infringed on its intellectual property) but then fully embraced the involvement of its large fan base in the development of new products. Large firms such as GSK or Google even mobilize their own employees as potential users of their products ("eat your own dog food").

Rather than trying to integrate external users within their own development processes, firms can now "go in the field" and interact online and/or "in vivo" with their users in order to sense, prototype, validate or refine potential solutions in multiple and evolving real-life contexts. This can involve sending people to customer premises in order to realize ethnographical observations. It can also involve creating fully dedicated spaces ("labs") where users and partners engage in real-life creation and experimentation. Finally, it involves more frequent tracking, mining and analyzing of what users are doing and saying online, for example, on social networks.

IDEO is an example of a firm that relies on careful "in situ" observations of users in order to uncover untapped value gaps. More recently, "Living Labs" are examples of initiatives aimed at interacting with users in "real-life" settings, and most big consumer goods firms buy and "mine" information from Facebook or Google regarding the behavior of their users.

So What?

Customer surveys, interviews, focus groups and trends analysis are necessary for many firms to sustain their competitive advantage. But more user-centric and embedded approaches and techniques must also be proactively and systematically implemented. These approaches, combined with powerful analytics, can allow firms to uncover untapped value gaps and new sources of competitive differentiation. Innovation managers must leave the comfort of their offices and "Get out of the building"!

4.4.3 Seeding New Ventures: Corporate Venture Capital

Since Christopher Columbus, kings and other powerful interests have always kept a close eye on what was going on at the edge of their current knowledge. They have sponsored explorers to keep them informed about the remaining "terra incognita" and preempt potential new discoveries. In the same way, getting involved with disruptive start-ups working at the edge of existing technologies can be a valuable source of innovation opportunities for incumbent firms. This can take the form of direct investments in start-ups through corporate venture capital ("CVC", Dushnitsky and Lenox 2006) initiatives, or through more indirect and informal initiatives such as start-up events.

Artificial intelligence, virtual reality, genetic engineering and blockchain technologies are examples of areas where incumbent firms are getting involved with entrepreneurial ventures in order to better understand "what is going on" and preempt potential opportunities.

Fertile Environments

The development of entrepreneurial equity markets and industrial revolutions in fields such as information technology and biotech have led many firms to invest corporate funds in independent external start-ups as a way to "get involved" in the revolutions.

This could be done directly by the parent firm, or indirectly through a dedicated fund managed by a separate financial operator. In the latter case, the firm can be a "passive" investor in an independent venture capital fund or an active partner in a dedicated fund, focusing on a specific sector, industry, technology and/or geographic area. The parent corporation can invest in the funds as an exclusive limited partner, or jointly with other corporations sharing an interest in the same sector or technology.

Such "seeding" approaches are particularly effective in dynamic industrial environments with weak or limited intellectual property protection, where capabilities are spread around multiple players and where a strong venture capital community exists.

Examples of CVC structures include ICT and pharmaceutical companies, such as IBM, Intel or Johnson & Johnson, but also more industrial firms, such as Siemens, Bayer or BASF, that invest in sectors such as energy and environmental care, automation and control, industrial and public infrastructure, biotechnology or nanotechnology.

Make Small Bets

Many CVC investments were initially seen as diversification and/or financial bets. Such investments allowed the parent company to support the development of its own technologies, platforms and ecosystems, including by investing in potential customers. It was also seen as a way to build options, by taking small stakes in emerging firms that had the potential to become successful businesses.

However, combining such strategic and financial objectives with the uncertainties of trying to "pick future winners" often proved costly, time-consuming and difficult to manage effectively. The intrinsic volatility of these CVC investments, the difficulties of attracting and incentivizing skilled fund managers and the challenges of interacting with start-ups led many firms to downsize or abandon many CVC initiatives.

Investing in start-ups at a very early stage (seed investing) is on average unprofitable, even for professional financial investors. It is therefore not surprising that most corporations that tried to do the same failed.

Learning from the Front Line

While making financial or strategic bets on new ventures remains a challenging approach, learning from them can still be a key opportunity. Interacting with disruptive ventures allows incumbents to maintain some technology and competitive intelligence in a relatively lean way. Incumbents can in particular gain exposure and access to emerging complementary and disruptive technologies or business models, allowing them to identify and monitor new opportunities such as market discontinuities or new dominant technological design.

Such "externalization" of R&D and/or marketing activities to new ventures also allows firms to identify new potential partners and create connections with them. Finally, being exposed to entrepreneurial ventures and investors can also act as a valuable "culture shock", helping firms to challenge their routines and processes in a fast-changing environment.

Dominant Internet players such as Google or Amazon are constantly monitoring the "start-up scene", to identify and preempt new threats or opportunities. Some financial institutions such as ING are also learning about the impact of the "Fintech" revolutions affecting them by getting directly involved with carefully selected ventures and investors.

On top of such equity investments, incumbent firms are also discovering other non-financial ways to get involved with start-ups, in order to identify potential partners, test concepts or gain new sources of inspiration. Conversely, start-ups are increasingly attracted by the scale, capabilities and reach of large corporations, whether as future partners or customers.

Corporate start-up "speed-dating" events include various types of "start-up fairs", such as the so-called corporate hackathons inspired by the software industry. While initially aimed mainly at trying to "pick winners" and accelerate their development, the events are increasingly seen more as external networking and learning opportunities for both start-ups and big firms. Opportunistic intermediaries now even offer big corporations full-fledged "venture tourism" packages in places such as Silicon Valley or Tel Aviv.

So What?

The increasing speed and breadth of technology evolutions make it impossible for firms to build capabilities that cover all the potentially relevant areas. Getting involved with carefully selected start-ups, either through direct investments or through cross-fertilization events, can be a cost-effective and fast way to develop market and technology intelligence in emerging industries.

4.5   Synthesis

Identify Attractive Innovation Opportunities: Key Insights

4.1. Identify the sources of innovations: beyond R&D

i. *R&D is not enough.* Significant R&D spending might be necessary for some firms to develop new technologies but certainly do not guarantee their innovation success. Identifying innovation opportunities requires finding new combinations of new or existing technologies with new or existing needs.

ii. The *triggers of innovation opportunities* which firms should proactively exploit include internal sources such as new knowledge, challenging routines and serendipity, and external sources such as changes in markets, industry and environmental constraints.

iii. *Sizeable innovation opportunities* do not pop up out of the blue. Organizations must invest time and resources to combine, integrate and mature innovation ideas into potential opportunities. They should also pursue already emerging opportunities and focus on scaling them up.

4.2. Foster organizational learning: beyond ideation

i. *Organizational learning* is about growing the intellectual capital of the firm across its people and teams, mobilizing problem-driven, opportunistic and systematic search behaviors.

ii. To uncover potential innovation opportunities, organizations and employees need to learn how to generate ideas by *thinking in new boxes*, challenging their prevailing mental models and "reinventing new wheels".

iii. *Employees* can be a rich source of learning and innovation provided that goals and expectations as well as coaching, selection, follow-up and feedback processes are carefully managed. Even in the best organizations most ideas end up being rejected. What matters most is what actually does happen the day *after* the "ideation" events.

iv. Organizations must continuously develop their technology base—manage R&D operations—but also know when to build *new technology platforms*—deploy an R&D strategy—as formerly irrelevant or emerging knowledge and skills become core or even distinctive.

4.3. Harvest and protect organizational knowledge assets

i. To identify attractive opportunities, organizations must systematically manage their explicit and implicit *knowledge assets*, finding ways to share, combine, disseminate and maintain this unique type of capital.

ii. Innovation opportunities can be sources of competitive differentiation provided that the intellectual capital they are based on is actively *protected* through a mix of secrecy, lock-in and/or intellectual property rights.

iii. Innovation opportunities can be leveraged if they are supported by effective *patenting* strategies, balancing protection and disclosure. Not everything that can be protected should be, and patents can and should foster rather than hinder collaboration.

4.4. Integrate external sources of knowledge

i. On top of their own R&D activities, product roadmaps and experience curves, firms must develop the ability to proactively *harvest technology (r)evolutions*. They must be able to absorb technology intelligence from external sources, both in their socioeconomic ecosystem and in the wider environment.

ii. On top of their traditional marketing approaches, firms need to develop user-centric ways to "pull" *untapped customer needs* and uncover value gaps. They need to combine powerful analytics with in situ and empathetic observations as well as lead users' interactions and involvement.

iii. Firms exposed to technology-intensive sectors (i.e. most firms) should invest time, money and resources to network and work with disruptive start-ups, in particular through *corporate venture capital* initiatives. This must allow them not only to create options and leverage their assets but also to effectively develop new competitive and technology intelligence.

Bibliography[1]

Cohen, W. M., & Levinthal, D. A. (1990). Absorptive capacity: A new perspective on learning and innovation. *Administrative Science Quarterly, 1990*, 128–152.

Drucker, P. (1985). The discipline of innovation. *Harvard Business Review, 63*(3), 65–72.

Du Plessis, M. (2007). The role of knowledge management in innovation. *Journal of Knowledge Management, 11*(4), 20–29.

Dushnitsky, G., & Lenox, M. J. (2006). When does corporate venture capital investment create firm value? *Journal of Business Venturing, 21*(6), 753–772.

Grant, R. M. (1996). Toward a knowledge-based theory of the firm. *Strategic Management Journal, 17*(S2), 109–122.

Greve, H. R. (2003). A behavioral theory of R&D expenditures and innovations: Evidence from shipbuilding. *Academy of Management Journal, 46*(6), 685–702.

Kaplan, S. (2008). Framing contests: Strategy making under uncertainty. *Organization Science, 19*(5), 729–752.

Nahapiet, J., & Ghoshal, S. (1998). Social capital, intellectual capital and the organizational advantage. *Academy of Management Review, 23*, 242–266.

Nonaka, I. (1994). A dynamic theory of organizational knowledge creation. *Organization Science, 5*(1), 14–37.

Pisano, G. P. (2006). Profiting from innovation and the intellectual property revolution. *Research Policy, 35*(8), 1122–1130.

Samuelson, P. A. (1954). The pure theory of public expenditure. *The Review of Economics and Statistics, 36*(4), 387–389.

Von Hippel, E. (1986). Lead users: A source of novel product concepts. *Management Science, 32*(7), 791–805.

---

[1] An extended bibliography is available at www.NavigatingInnovation.org

5

Develop a Balanced Portfolio of Business Models

Most innovation ideas do not lead to robust business opportunities, and most organizations try to pursue more business opportunities than they can actually successfully handle. In order to effectively manage innovation, firms must therefore select the opportunities for which a convincing business model can be designed and which collectively form a balanced and consistent portfolio.

The fourth innovation management challenge is therefore to focus on the right portfolio of innovative business models: asking the right questions, designing competitive business models and mobilizing the right resources, valuing those business models and combining them to build a consistent and balanced portfolio (Fig. 5.1).

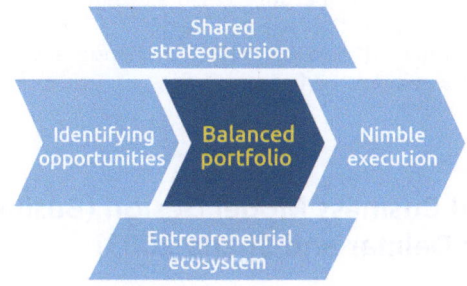

5.1 Asking the right questions: business model design
5.2 Designing competitive business models: why and what
5.3 Mobilizing the right resources: who and how much
5.4 Valuing innovative business models
5.5 Building a consistent and balanced innovation portfolio
5.6 Synthesis

Fig. 5.1 Developing a balanced portfolio of business models

© The Author(s) 2018
B. Gailly, *Navigating Innovation*, https://doi.org/10.1007/978-3-319-77191-5_5

5.1   Business Model Design: Asking the Right Questions

Designing or improving a successful business model based on an identified innovation opportunity entails learning how to (better) answer, in a credible and consistent way, a set of key questions (Fig. 5.2).

While various approaches, templates and canvases have been proposed and can be used to answer this set of key questions, what matters is whether a convincing "story"—rather than a plan—can be developed over time, validated and sold to whoever needs to be mobilized.

Key Insights

i. Designing or improving a *successful business model*, business planning, means on the one hand building and validating a credible story regarding how specific resources could be (better) mobilized to solve a specific problem and on the other hand selling that story to the relevant internal and external stakeholders.

ii. A business model will be convincing if it addresses in a consistent way four *key questions*: (1) *Why* is there a problem and why are we well positioned to solve it? (2) *What* exactly could be sold to whom and how? (3) *Who* needs to be mobilized? (4) *How much* is at stake?

iii. Finding new ways to address the why, what, who and how much key questions around the same innovation opportunity can allow managers to *design innovative business models*.

iv. Successful entrepreneurs and investors *do not plan to fail*. They prioritize the "why?" and "who?" key questions when assessing an innovative business opportunity. They know that the technology specifications ("what?") and financial spreadsheets ("how much?") will change and will need a lot of time and further effort to be fixed.

5.1.1 Successful Business Model Design (Business Planning; Delmar and Shane 2003)

This Is Not a Plan

The objective of designing a successful business model (Teece 2010) is not to plan what will exactly happen, which is foolish when dealing with emerging innovation opportunities. Managing this process is therefore often challenging for engineers or scientists, whose reputation tends to be correlated with their accuracy.

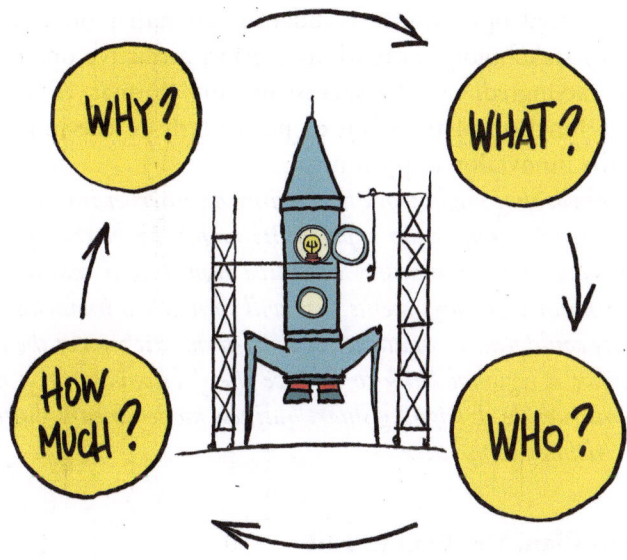

Fig. 5.2 Business model design

The objective of designing a successful business model is on the one hand to develop in an iterative way a credible story regarding what could happen and on the other hand to validate whether value could be created and captured according to that story. A business model is therefore a set of assumptions, regarding how an organization could perform over time and mobilize various stakeholders in order to create and capture value.

The first business model for Amazon probably did not plan to offer cloud computing and outsourcing services or to design and sell new electronic readers. It was however still successful, because it convinced early investors and employees to commit significant time and resources in order to pursue an innovative opportunity.

Pick Your Canvas

Designing the right business model "story" means identifying the story's target audience and the story's objective. The audience of a business model design can include internal stakeholders such as colleagues and potential recruits as well as external stakeholders such as investors, specialized suppliers or corporate executives. They all have different perceptions and expectations regarding what a successful business model should be, which must be taken into account.

Depending on the target audience and context, the objective of a business model design can therefore be to focus on reaching a consensus regarding the value of a potential opportunity (validating) or rather on convincing and mobilizing key stakeholders around an a priori attractive opportunity (selling). Finally, the ingredients of a successful story (format, content, delivery) will also be a function of the context, particularly of the industry and the maturity of the innovation opportunity.

A specific canvas that might work to convince an internet investor regarding an emerging B2C service innovation opportunity might not be the right one to convince a pharmaceutical or industry executive regarding a mature B2B process innovation. Similarly, the arguments that will convince a financial investor that an opportunity could lead to an attractive short-term exit are not the same as those used to convince a regional bank to finance the development of a new capital-intensive infrastructure. Designing smart business models means much more than just filling predefined templates.

If You Fail to Plan, You Plan to Fail

Understanding business planning as a design and storytelling exercise also highlights the fact that the process often matters as much as the outcome. The hypotheses supporting a given business model might quickly end up being wrong or outdated. However, the process itself, of validating (or not) through multiple iterations and with multiple stakeholders the various scenarios, and checking the consistency of the various dimensions of a business model, will ultimately reinforce the robustness of the business opportunity. Conversely, trying unsuccessfully to design a business model can help the organization to quickly stop exploring opportunities when no credible business model can be designed around them.

While the future is always unpredictable, shaping and framing various future paths makes one better prepared to face whatever could happen. Conversely, if it appears impossible to build any credible story according to which an innovation could create value, it probably means the opportunity should be abandoned. If it is too good to be true, it is probably false.

While both faced completely unpredictable and utterly challenging circumstances, the differences between Roald Amundsen's success and Robert Falcon Scott's failure in their race toward the South Pole can be linked to their preparation and planning decisions. Scott's initially bigger team reached the South Pole more than a month after Amundsen and died during the return journey not only because of bad weather. They failed also because they made the wrong initial assumptions and choices, such as relying on ponies and motorized sledges rather than dogs.

So What?

Good business planning for innovation is often wrongly assumed to mean relying on standard approaches and templates in order to try to accurately predict what will actually happen. But designing successful business models based on innovation opportunities is not about planning the future. It is about building consensus and progressively coming up with a consistent and credible story regarding what could happen and how value could be created and captured.

5.1.2 The Key Questions: Why (Us)? What? Who? How Much?

An innovation opportunity will be attractive from a business point of view if on the one hand a competitive market opportunity can be identified (Why and What) and on the other hand the resources required to capture the opportunity can be mobilized in an effective way (Who and How Much).

Why and What: The Opportunity

Competitive business is about both addressing customer needs better than others and managing to sell and deliver to the customers an actual value proposition. The first key questions regarding an innovation opportunity are therefore whether it could allow the firm to solve a problem more effectively than existing alternatives ("why/why us") and whether the value created as a result could actually be captured ("what").

In particular, an identified market need and a validated technology are often necessary but never sufficient conditions for an innovation opportunity to be attractive. What also matters is both whether the firm is better positioned than others to address that market need, and whether an actual value proposition can be derived from the validated technology.

Countless inventors identified new ways to solve customer problems but failed to capture the resulting business benefits, because others were better positioned and quickly copied them. Similarly, there are technological solutions to many healthcare problems, such as water sanitation or disease prevention, that have not yet been turned into a value proposition anybody would actually pay for. As other examples, one of the key differences between Google's success and Twitter's struggle is that while both address significant market needs, the former has implemented an effective way to monetize its technology (through advertising) and the latter has not.

Who and How Much: The Resources

Having identified a competitive market opportunity and an actual way to monetize that opportunity is not enough. What is also needed to decide whether an innovation opportunity is attractive for a firm is to assess whether the firm could effectively mobilize the required human and financial resources.

From a human resource point of view, this means assessing whether the required talent, expertise and partners can be mobilized, and whether an adequate governance structure—a way of working together—can be set up. An opportunity is worthless if there is nobody available to capture it. The innovation graveyard is full of projects that failed because the right people could not be mobilized or because internal conflicts paralyzed their management.

The quality of the project team is known by professional investors to be a key success factor for innovation projects. However, too often corporations will first decide to launch a project and only then try to staff it with whoever happens to be available.

From a financial point of view, this means trying to assess whether the project is ultimately "worth it", taking into account the opportunity cost of the required resources, the size of the potential and the risks and time horizons. While putting numbers on an emerging innovation project is sometimes a tricky endeavor, it often provides a much needed early "reality check" in terms of potential profitability.

The financials of an emerging innovation project are often the least reliable facts known about that project. But in business, numbers often speak louder than words. Even for not-for-profit organizations, assessing whether an innovation opportunity represents an effective use of its resources should be a key issue.

Addressing the Weakest Links

Probably the worst way to address the four key questions underlying the design of an innovative business model (why, what, who, how much) is sequentially and separately, as distinct phases of a linear process. The design of an innovative business model should be an iterative and integrated process. What matters is to converge toward a consistent set of answers, or to conclude that no convincing "story" can be designed. This means identifying a project's weakest points at each stage, and then addressing them, adjusting the various elements of the business model along the way.

Most of the successful start-ups we know today actually made several significant adjustments to their business model (they "pivoted") as they evolved from an initial innovation idea toward a sizeable and competitive business based on a competitive market opportunity, a realistic value proposition, a governance structure and a profitable outlook.

So What?

The design of a potentially successful business model around an innovation opportunity must be managed as an iterative and integrated process, focused both on the opportunity (why/why us and what) and the resources (who and how much). Too often firms focus excessively on trying to definitively address some aspects of an innovation opportunity, such as expected returns or technical feasibility. They end up being exactly wrong, rather than trying to converge toward what could be approximately right.

5.1.3 Designing Innovative Business Models (Foss and Saebi 2017)

Designing a potentially successful business model requires finding at least one consistent set of answers to the "why", "what", "who" and "how much" questions. But doing so also creates room for new types of answers, allowing various innovative business models to be designed around the same innovation opportunity.

In many cases, designing the right business model is actually more important than having identified the right innovation opportunity.

An innovator who conceived a new type of electric battery could license the related intellectual property rights, provide advisory services to manufacturers or energy providers, manufacture pieces of equipment or full installations, become a back-up energy provider or use the produced energy to sell something else. These are all potential business models, which could be designed around the same new technology, but with completely different implications in terms of resources, risks and potential.

New Whys and New Whats

The first way to design an innovative business model is to identify new business opportunities built around the same innovation. This can include addressing new market needs with the same innovation, leveraging new sources of competitive advantage, designing new types of value proposition around that innovation or mobilizing new types of value chains. In all cases the underlying innovation can be the same but the businesses built around it might end up being completely different.

An innovation such as cloud computing has seen innovative business models emerge around new market needs, such as "green" and scalable IT infrastructures for start-ups; new sources of competitive advantage such as artificial intelligence capabilities or regulatory arbitrage (e.g. regarding privacy and data protection laws); new value propositions, such as encryption protections and pay-per-use or freemium models; and new value chains roles, such as advisory services or integrators. These are all different and innovative business models, built around the same initial innovation opportunity, but with completely different requirements and potential.

New Whos and New How Muches

The second way to design an innovative business model around a given innovation opportunity is to find new ways to mobilize the required human and financial resources. This can include new ways of working as a firm or new forms of governance, built around new types of organizations, partners or intermediaries. This can also include new ways to raise funds and mitigate risks as well as new ways to structure, manage and reward stakeholders.

Business model innovations built around new ways to mobilize resources include decentralized peer-to-peer platforms such as Uber, open-source organizations such as Wikipedia as well as crowdfunding intermediaries. They also include the new types of shareholding structures and agreements built around family conglomerates or the "layered" voting rights of Facebook and other internet champions.

While many of those business models, such as crowdfunding, have actually existed for decades or even centuries, the wide and relatively low-cost availability of internet-based services and devices has given them new life, as they made their implementation and adoption much easier than before.

So What?

The business model is what can turn an innovation opportunity into an actual business. Innovation managers should therefore invest as much attention and resources in its design as they have invested in the development of the innovation itself. They should also explore opportunities to develop new business models around existing innovations, including around innovations initially developed by others.

5.1.4 Do Not Plan to Fail

Designing a successful business model requires addressing in a consistent way the four questions of why/why us (strategy), what (marketing and operations), who (human resources and governance) and how much (financing and valuation). While the ultimate success or failure of the resulting business will remain strongly affected by factors beyond the control of its managers, there are common pitfalls that can be avoided. The first pitfall regards focusing early on the wrong issues and the second regards overoptimistic planning.

The common myth about successful start-ups is that they were overnight successes, based on brilliant revolutionary ideas. But experienced investors and seasoned entrepreneurs have learned what to focus on at the early stage of an innovation project (the right team and the right market opportunity) and how much crossing the chasm between an idea and a business is a long journey, full of twists and turns. And yet even these experts often get things wrong.

Early Focus: The Right Persons at the Right Place and the Right Time

Most successful businesses end up selling something completely different from what they initially had in mind. On top of that, their initial financial predictions had the accuracy of an astrologist. Yet too often managers assessing an early innovation opportunity will excessively focus on such technical specifications and financial numbers. In other words, they give the most importance to what is initially the least reliable.

Experienced innovators know that while early prototypes ("what?") and financials ("how much?") can be useful, what matters most initially is to be able to mobilize the right team ("who?") and to be in the position to exploit the right market opportunity ("why/why us?"). While there are cases where the initial team and/or the initial market opportunity were completely restructured, they remain the most robust parts of the majority of early business plans.

Facebook, Airbnb, YouTube, Google and many others started with value propositions and financial expectations that were often widely off the mark. But they all had a great initial set of people and an attractive market opportunity. In contrast, many failed businesses focus too much on technology and fancy financial expectations and have no understanding of how to capture markets in a competitive way.

Escaping the Planning Fallacy

The second common pitfall of business model design is an overoptimistic assessment of the steps and time needed to turn an idea into an actual business. This can be linked on the one hand to the intrinsic optimism of most entrepreneurs and innovators—if they knew how difficult it was they probably would not try it—and on the other hand to the inherent resistance to change of people and markets—newness takes time, and nearly every time much more than expected.

As a consequence, structuring, setting up and scaling up the necessary operations while remaining competitive will in most cases take more time and be more difficult than expected. Similarly, convincing users to adopt and/or pay for an innovative value proposition will often take more time and effort than planned. Even organizations that know that most of their past innovation projects took more time and more effort than planned for keep being overoptimistic in their planning of new innovative projects.

A perverse consequence of this overoptimism is that many ventures that actually had a significant value creation potential end up being abandoned because they ran out of cash or energy before being able to demonstrate that potential. Some can still try to raise new funds or get new resources and budgets but are often much less convincing and in a much weaker bargaining position.

Even in a digital world, turning an idea into a successful sizeable business often takes five to ten years and has the odds of a lottery ticket. Some large corporations are keenly aware of the hurdles and have therefore mastered the art of "adopting teenagers rather than making babies". They let many small or medium-sized firms exhaust their time and energy developing new ideas and then pick the winners, buy and scale them up.

So What?

Designing successful business models remains an art more than a science and is still a very unreliable process. But experienced investors, entrepreneurs and innovators know what to focus upon initially: teams and market opportunity rather than financials and technology. They also know that going to market will require more insights, time and effort than most people plan for.

5.2 Designing Competitive Business Models: Why and What?

Designing a successful business model based on an innovation requires identifying a business opportunity which is attractive from both a strategic and operational point of view.

From a strategic point of view, this means on the one hand assessing whether the business model targets a market with sufficient potential—"why?"—and on the other hand whether the firm has reasons to believe that it could be better placed than current and future competitors to capture that potential—"why us?" (Fig. 5.3)

From an operational point of view, this means assessing whether the opportunity can actually be monetized in some ways—"what could be sold?"—and whether the activities needed to sustain the business model could actually be implemented—"what could be done?"

Key Insights

i. The first (obvious but still too often forgotten) strategic challenge of a successful business model is to *address somebody's problem*. It must demonstrate that potentially enough people out there have a problem and that the problem is painful enough for them to be ready to both adopt and pay for a new solution.

ii. The second strategic challenge of a successful business model is to be *better positioned than others*. The innovators and their organization must demonstrate that they could address a relevant problem or meet a need better than the available alternatives, based on their scope as well as their scale, unique assets and/or agility.

iii. On the other hand, the first operational challenge of a successful business model is to conceive an initial *value proposition*, which some people will actually be ready and able to find, adopt and pay for. What will be on the first invoice?

iv. The second—and too often forgotten—operational challenge of a successful business model is to set up, integrate and scale up over time a *competitive value chain*, including the right design, operations, client management and support activities.

5.2.1 Strategy: Why? Addressing Somebody's Problem

Probably the most basic but too often forgotten question regarding an innovation opportunity is: "Does it solve somebody's big problem?" More specifically,

Fig. 5.3 The main sources of business models' competitive advantages

is there a significant number of people for whom the perceived benefits of adopting the innovation could be significantly greater than the perceived cost of disrupting their routine?

Houston, Who Has a Problem?

Many innovators end up falling in love with their innovative product or technology. As a consequence, they become blind to the fact that most other people actually do not care, or at least do not care enough to pay for it. They fail to gain a good understanding of who actually worries about a significant problem that the innovation could potentially address.

Many innovative payment solutions struggle to gain market share because the users they target do not actually perceive that they have a "payment problem", or because they worry about other things (privacy, security, convenience, acceptance, etc.) which for them are more important than adopting the latest gadgets.

Designing a successful business model around an innovation opportunity therefore requires gaining an in-depth understanding of who could need what the innovation can offer, who could pay for it and who actually could make and influence the purchase decision.

This entails learning about, tracking down and continuously analyzing the characteristics—position, social profile, aspirations and behaviors—of the people most likely to adopt the innovation and/or motivate others to adopt it.

It also entails gathering evidence that there might be enough such people to justify an actual business opportunity.

Too often prospective innovators rely too much on hypothetical markets or abstract organizations as justification of their potential. But what matters is whether there are real-life people out there who need a solution. A business model that identifies and names actual potential adopters is often much more convincing than a spreadsheet that relies on fancy quantitative predictions. Even in B2B markets, where the target customers are organizations, innovation managers need to identify the actual human beings behind these organizations, those who could approve purchase decisions and sign invoices.

Empathic Design

Steve Jobs supposedly said that "It is not the customers' job to know what they want." Having identified a set of potential—real-life—adopters, the next challenge is therefore to gain a good understanding of the exact problem they face, and to do so from their prospective. Most purchasing or adoption decisions are driven not only by technical features and rational requests but also by subjective emotions, perceptions and values. Individual consumers but also corporate managers will be affected by a variety of buying criteria, framed by their personal perceptions and viewpoints. And if the innovation addresses a problem that might be real but that they are not aware of, or that has a low priority, they will most probably not adopt it.

In the past, many industrial firms (e.g. automotive or aeronautics suppliers) could wait for explicit specifications and requests for proposals from their customers in order to learn what exactly they needed to provide. But today, even in B2B markets, the challenge is increasingly to be able to anticipate future needs and to proactively identify problems customers do not know they are or will be facing.

In order to design a successful business model around an innovation opportunity, it is therefore critical to gain a good understanding of why some people could adopt an innovation, the perceived needs they aim to fulfill and the conscious and unconscious decision criteria that influence adoption. The actual motives to adopt (or not) an innovation can end up being very different from what the innovators initially saw as the most important features of their innovation.

Many innovative preventive technologies have been developed to protect people from the electromagnetic fields of a mobile phone or from nosocomial bacterial infections. But many were rejected by telecom operators and hospitals because they saw them not as potentially life-saving products but as potential admissions of future liabilities and as public relations risks ("You knew that there was a problem and you waited so long to deal with it!"). Similarly, the motives behind the investments in a new corporate headquarters or the purchase of a new high-end car can be more emotional than rational.

Finally, the challenge is to understand the needs and expectations not only of the target end-user but also of all the key intermediaries and influencers. Only if the needs and expectations of most of them are met in some ways is there a chance for an innovation to flourish.

An expensive equipment upgrade aimed at decreasing maintenance costs might delight end-users but maybe not the installers the innovator needs to convince to promote it. Similarly, a vaccine against mumps with rare but dangerous potential side-effects might delight parents and toddlers but might create a dilemma for some pediatricians for which the disease is an important source of patients.

So What?

While quantitative predictions regarding the potential market for an innovation are known to be very unreliable, gaining an in-depth qualitative understanding of who has a problem and its exact nature is a prerequisite for most successful business models. Innovators must focus less on crunching numbers from third-party market studies and more on "getting out of the building" in order to experience the circumstances and problems actually faced by potential real-life users and customers.

5.2.2 Strategy: Why Us? Being Better Positioned Than Others

Having identified a sizeable market need that could be addressed by an innovation opportunity is not enough. What also matters is whether the opportunity could fit with the strategy of the organization and in particular whether the firm could be better positioned than others to capture the innovation opportunity. Pursuing irrelevant or losing innovation opportunities is a waste of corporate resources.

Being on the Corporate Agenda

In most cases, pursuing a given innovation opportunity is only one of the many options a firm has regarding the allocation of its resources. Moreover, it is often not one of the most appealing options in terms of short-term financial rewards. Making sure that the innovation opportunity fits with the strategic priority of the firm is therefore a prerequisite if one hopes to be able to efficiently mobilize corporate resources to capture the opportunity.

The leading credit card players had been aware for many years of the threat of new online players such as PayPal. However, their traditional "plastic" business was growing so quickly and profitably that it was nearly impossible for such disruptive opportunities to gain traction. Only when new regulatory constraints started to threaten their core business was it possible for these projects to gain a place on the corporate agenda.

Designing a successful business model based on an innovation opportunity therefore entails assessing whether it could fit with the technical and organizational resources, the corporate environment and the strategic priorities of the organization. Trying to capture a sizeable opportunity "on top of" existing business activities and budgets or as a "nice to have" low-priority project is a recipe for failure. If a disruptive project cannot find a legitimate place on the executive agenda, it is very unlikely to succeed.

Most innovation managers have learned during a change or project management training that in theory they need the strong support of a sponsor if they want to develop a disruptive innovation. But too often they fail to invest enough time, attention and resources in meeting that need, focusing rather on technical or financial issues.

Organizations try too often to pursue innovation opportunities that actually lie outside the (sometimes implicit) strategic agenda of their management. And pursuing an opportunity that actually could work but is not considered strategic by the organization is probably one of the most frustrating experiences for an innovation manager.

As much as half of the successful start-ups in a given industry are launched by people who used to be corporate managers but decided to leave their employers because of their perceived lack of support.

Play to Win

Having identified a sizeable innovation opportunity that could fit within the corporate strategic agenda is not enough. What also matters is whether the firm trying to capture that opportunity is better positioned than others to appropriate it. If an innovation is attractive, then existing and new competitors will try to copy it. And if the firm cannot be better than them, they will win.

Countless traditional firms have tried to develop "digital" opportunities because it was perceived as a strategic priority. Most have failed, because they did not have the skills, resources and capabilities required to master the key success factors of internet businesses (visibility, user experience, logistics, etc.) and therefore could not develop competitive online value propositions. The same phenomenon can be observed in many industrial firms that try to capture "servitization" opportunities, without the adequate skills, infrastructure and resources.

While the specific ways to be "better than others" are obviously functions of the industries and markets considered, there are three main sources of competitive advantage that should at least be considered when designing a potential business model: scale, unique assets and agility (adapted from Dierickx and Cool 1989).

Innovation managers must therefore assess whether a combination of scale, unique assets and agility could be leveraged in order to capture an innovation opportunity and prevent competitive imitation, substitution or hold-ups. In each case the question to consider is whether the firm has or could develop the capabilities to capture the innovation opportunity better than existing or potential competitors could.

"Scale" relates to the ability to capture economic profit from market power. The biggest player in an industry enjoys bargaining power and economies of scale and scope others cannot achieve.

"Unique assets" relate to the rents generated by the critical and scarce resources that competitors cannot easily access or develop, such as patents, facilities or unique talent.

"Agility" relates to the ability of some organizations to identify and capture opportunities more quickly than others, capturing values before competitors can react.

Many online retailers are minnows compared to Walmart and Carrefour, which could leverage their scale to develop very competitive online businesses. The brands of Apple or Mercedes are unique assets that Samsung and Huawei or Lexus still have difficulties competing with. Finally, L'Oréal has successfully identified and captured multiple market opportunities in the personal care sector, skimming off profits before other bigger or more cost-effective players could react.

So What?

Innovation managers who design innovative business models must carefully consider on the one hand whether the opportunity could find a high-priority place on the corporate agenda and on the other hand how much the firm could leverage or develop a combination of scale, unique assets and agility in order to out-compete available alternatives. Addressing a sizeable need ("why?") is not enough; it has to be the right need for the corporation to pursue ("why us?").

5.2.3 Operations: What Could Be Sold? Designing the Value Proposition

Experienced sales and marketing managers know very well that people will actually adopt and pay for a product or service for reasons which go well beyond the technical features or specifications of that product or service. Designing a successful business model around an innovation opportunity therefore means much more than designing a high-performance device or software. It means also thinking about how an attractive value proposition could be designed around this device or software.

Innovation managers must therefore consider not only the "core" attributes of the innovation, how it could be a solution to a specific problem, but also how it will meet and sometimes exceed users' expectations regarding all the basic, expected and potential features they will enjoy.

People buying a new electric car expect not only a "green" car and a charging infrastructure. They also expect regulatory compliance, safety, reliability, comfort, a reliable and easily accessible aftersales and maintenance network, a high resale value and many other features.

What Will Be on the First Invoice?

One way to characterize the key elements of a potentially convincing value proposition is simply to imagine what will be written on the first invoice sent to the first adopter. Checking that anything could actually be invoiced one day is a simple test, but it often immediately raises tough questions.

Designing such a hypothetical invoice means of course clarifying what the product or service will feature. But it also means thinking about its pricing, packaging and distribution, all issues too often neglected by innovation managers. Finally, it means also thinking about the communications and services surrounding the offer before, during and after the purchase, both offline and online. And this needs to be done considering not only the end-users, but also all the key intermediaries and influencers.

An innovative car part sold in bulk to car manufacturers three months before delivery will have a very different profitability than exactly the same car part sold as individual orders to repair shops and delivered in 24 hours, with packaging designed to optimize handling. Similarly, a medical device will be adopted not only because of its health impact but also because of its regulatory compliance, its cost for the hospital, the patient and the healthcare system, its ease of installation, use and maintenance for the hospital as well as for its perceived convenience and risk for doctors and patients.

Seeing Is Believing

An important implication of the differences between a technology and a value proposition is the importance of prototypes, of demonstrating what people could actually pay for. Innovation managers must therefore find creative ways—such as mock-ups or virtual prototypes—to actually show as early as possible what the innovation could look like. A pitch with a demo is often much more convincing than a PowerPoint full of charts and tables.

A key step in the adoption of ICT technologies was the first live demonstration of a user-friendly computer interface by Douglas Engelbart in 1968, including multiple windows, hypertext and file linking, graphics, efficient navigation, video conferencing, a computer mouse and word processing.

So What?

Innovation managers must always remember that people do not adopt innovative technologies, they buy innovative solutions. This means that designing and showcasing a potential full value proposition around an innovation must mobilize as much attention and effort as designing the innovation itself does. The success of an innovative offer will often be driven as much by its marketing (pricing, packaging, distribution, promotion and complementary services, online and offline) as by the performance of the underlying technology. Innovation managers should therefore invest as much attention in how they will "go to market" as they do in the specifications of their technology.

5.2.4 Operations: What Could Be Done? Assembling a Competitive Value Chain

Having designed an appealing value proposition around an innovation, the second operational issue regarding the design of a successful business model relates to the actual implementation of the activities necessary to deliver and extract value. Many steps need to be completed to turn raw material or data into consumer value in an innovative way. They include design, operations, customer management and support activities that need to be effectively implemented and scaled up. A successful business model will exist only if a potential way to make these steps and activities happen can be found and allow the firm to deliver and extract sufficient value.

Designing an innovative process to turn organic waste into energy is one thing. Implementing all the activities required to process the waste upstream, distribute and sell the energy downstream, and finance, sell, install, maintain and operate the necessary equipment is another. The former is an innovative technology; the latter is an innovative business model.

Find Your Niche

The first step regarding the design of an innovative business model is to define the potential position that business model could take within the ecosystem, anywhere from raw inputs to end-user value. This means first picking the right new or existing industry value chain. This also means identifying within that value chain the best place for the firm. This choice should be made based on the one hand on the platform of unique capabilities the firm can leverage and on the other hand on its relative bargaining power vis-à-vis the various suppliers, complementors and customers it will face in its ecosystem.

While consumer brands make online retailers such as Amazon famous, selling innovative products or services to end-consumers is definitely not the only way to deliver and extract value out of an innovation opportunity. The most profitable players in an industry—at least in relative terms—are often unknown niche B2B actors specializing in a specific but important stage of the industry value chain, rather than the more glamorous B2C players. As an example, most people do not know Cargill or Qualcomm, although they probably use their product or service every day and both are huge and quite successful firms.

Amateurs Worry About Strategy, Professionals About Logistics

Whatever the potential attractiveness of the selected innovation opportunity, designing a successful business model also means thinking about how activities for delivering and extracting value could be effectively implemented and scaled up. Innovativeness is not a substitute for operational excellence.

From a mom-and-pop shop to a large service organization, the activities necessary for delivering and extracting value will always include design, operations, customer management and support activities, each of which needs to be managed effectively for the business model to be successful. Failing to deliver one of these activities means in most cases failing as a business, regardless of the innovativeness of the underlying opportunity.

The differences in performance between many "innovative" internet businesses often relate much more to their operational effectiveness than to the smart design of their product or service. Similarly, the person chosen at Apple to succeed Jobs was not the "chief creativity officer" but Tim Cook, its chief operations officer.

"Design" activities relate to the definition and continuous improvement of the value propositions developed by the business. These activities are often managed by departments such as R&D and/or marketing. They must focus in particular on the quality and competitive differentiation of the offer and are critical in, for example, "premium" industries such as luxury consumer goods.

"Operations" activities relate to the actual transformation of the inputs mobilized by the firm into the outputs necessary for delivering its value proposition. These activities are often managed by departments such as manufacturing, procurement and IT. They must focus on key performance indicators such as cost, speed and flexibility and are critical in, for example, commodity industries.

"Customer management" activities relate to the management of all interactions with potential and existing customers before, during and after the actual purchase decision. These activities are often managed by departments such as advertising, sales and customer service. They must focus in particular on key performance indicators such as perceived quality, trust and convenience and are critical in "relationship-driven" industries such as retail or private banking.

Finally, "support" activities relate to all the activities required for the first three sets of activities—design, operations, customer management—to be managed and scaled up smoothly. These activities are often managed by corporate support departments such as finance, human resources, regulatory, facility management or communications and focus on the satisfaction of their internal customers and external stakeholders. They are critical in particular in tightly regulated industries such as banking, aeronautics and healthcare.

Many traditional industrial companies found out the hard way that the key challenges of the digital economy or of the "servitization" revolutions were operational and not only strategic, as they had to deal with unexpected logistics, data management, human resource management or customer services issues.

Of course innovation managers do not necessarily need to create full corporate departments to support all these activities, especially at the initial stage of an opportunity. What matters is that a potentially smart and effective way to get these things done can be identified.

So What?

Innovation managers must focus not only on designing innovative value propositions targeting end-user needs. They must also think about which B2B or B2C role the firm could play in the delivery of value propositions and how to effectively and profitably implement and scale up the activities required to fulfill that role. Innovation is 1% imagination and 99% operational excellence.

5.3   Mobilizing the Right Resources: Who and How Much?

Designing an innovative business model requires not only identifying a way to capture a competitive and attractive business opportunity. It also requires making sure that the key resources needed to capture that opportunity could be mobilized (Haynie et al. 2009).

In particular, the key people—sponsors, managers, staff and partners—and the key financial resources must be identified, acquired and managed effectively (Fig. 5.4).

This means making sure that on the one hand the right people and structures can be mobilized and work together and on the other hand that the necessary financial means can be raised over time.

> **Key Insights**
>
> i. Probably the most important but least understood aspect of a successful innovative business model is the identification and mobilization of the required *entrepreneurial talent and expertise*. Too often firms decide first to launch a project and then try to staff and link it with whoever is available, that is, in many cases not the right people or partners.
> ii. The hidden secret behind the failures of many innovative ventures is not bad technologies but bad *governance*, with the wrong people taking the wrong decisions or failing to take any decisions at all.
> iii. The easiest way to waste an innovation opportunity is to underestimate the *financial resources* required over time to support the launch and growth of a sustainable business model. Great innovations are never overnight successes.

Fig. 5.4 Mobilizing the right resources

5.3.1 Entrepreneurial Talent and Expertise: Who Does What?

Having identified a potential competitive market opportunity around an innovation means only that the glass is half-full. What still needs to be done in order to design a successful business model is to consider the resources needed to capture that opportunity and where and how to find them. This is particularly the case for the most critical resource for an innovative venture: people.

Too often, budding entrepreneurs present themselves to potential investors on their own, forgetting that the quality of the team is one of the most important decision criteria for an investor. Similarly, corporations that launch, for example, employee-driven ideation processes might identify innovation opportunities but fail to correctly anticipate and address the issue of "who is actually going to do it?" As another example, during its first 20 years of existence the market value of Amazon has been closely correlated with its number of employees rather than its profitability.

Build a Winning Team

Managers or investors who consider an innovative business model must assess whether the talent, time and energy required to set up and develop the necessary activities can be mobilized at short notice and in a credible way. They need to consider not only the qualities of the potential project champion (if any) but also whether they can assemble an organization and a network of partners with the right skills and attitude. This includes in particular people with the ability to lead the various required departments, such as finance, technology, human resources, operations and strategy.

A convincing business model will therefore include a balanced team, with the right level of leadership and networking skills, industry and market expertise, track record and credentials, and finally the relevant functional capabilities needed to manage the design, operations, customer management and support activities.

A convincing business plan presentation should include a presentation of the current and future team members as well as current and future partners, with references, affiliation track records and levels of involvement. A good team with a bad idea is said to be much better than a bad team—or no team at all—with a good idea. And who is already on board can also provide a very strong signal regarding the credibility of the innovation opportunity.

Do Not Walk Alone

While defending an innovation project as a lone wolf is probably the first mistake innovation rookies make, the second is probably wanting to manage everything themselves. The diversity and complexity of the various tasks required to set up and scale up an innovative venture mean that in most cases they cannot be found within a single person or team. The ability to delegate and outsource is a key issue for any manager, but it is a critical one for innovation managers.

The ability to outsource many infrastructure issues, such as facilities (e.g. to incubators) or IT (e.g. to cloud computing providers), has been a key accelerating factor of many internet start-ups. Similarly, innovative ecosystems such as Silicon Valley provide entrepreneurs with easy access to specialized experts to facilitate their development, such as headhunters, accountants, public relations managers or lawyers.

How to assemble and manage a potential "constellation" of subcontractors, experts and partners, today and tomorrow, is therefore an issue that needs to be addressed in a convincing way when designing an innovative business model. This is particularly the case for corporate ventures, for which the ability to rely on the expertise and resources available within the parent corporation is both a key differentiating factor and an operational challenge.

Corporate ventures that fail to effectively benefit from the resources of their parent corporations end up combining both the disadvantage of being stand-alone— limited resources—and of being part of a large structure—limited autonomy.

So What?

The ability to demonstrate that the right talent, people and partners could be mobilized in order to capture an innovation opportunity is both one of the most important and one of most often forgotten issues when designing an innovative business model. Conversely, the ability to convince colleagues, recruits and partners to commit to a potential innovation opportunity is one of the key skills of successful innovation managers.

5.3.2 Governance: Who Decides What?

Having identified the potential talents and partners which could be mobilized to capture an innovation opportunity, the next challenge is to decide and agree on the governance structure to efficiently mobilize them. The governance of a new venture can be broadly defined as an explicit or implicit consensus regarding who should know what and who should decide what. It is—to borrow Peter Drucker's words—what "converts a mob into an organisation and human effort into performance". Governance of course needs to, at minimum, meet regulatory requirements, but it entails much more than legal paperwork or fiscal engineering.

One frequent obstacle to the growth of a new venture is the inability and/or unwillingness of the initial founder(s) to share power and responsibilities with the new stakeholders the venture mobilizes over time, such as investors, partners or high-potential recruits.

There are many possible governance structures that can be considered, with various levels of centralization, hierarchy, autonomy, individual freedom and specialization. The best one for a given innovative business model will be a function of the people involved, the opportunity, the socioeconomic and institutional context and the underlying cultural values and norms.

There are, however, two issues that should always be addressed for an innovative venture: the balance between strategy and execution and the level of autonomy of the venture vis-à-vis its main stakeholders.

Think Globally (Strategy) and Act Locally (Execution)

One of the key challenges regarding decision-making in an innovative venture is the need to combine operational effectiveness and strategic vision while facing growth, time pressures, ambiguous circumstances and scarce resources. Designing a convincing business model therefore means addressing how the governance structure of the venture will be designed to effectively handle these requirements, and in a scalable way. This means identifying who will over time on the one hand assume the responsibility of maintaining a consistent strategic vision and on the other hand provide the expertise and skills to effectively manage operations.

The "strategic vision" role can be assumed by, for example, a formal board or steering committee or by a more informal set of advisors and sponsors. The "operational support" role can be provided by specialized subcontractors and partners or, in the case of a corporate venture, by functional experts employed by the parent company.

In all cases, the strengths of the organization and its ability to implement a successful business model will be driven not only by the quality of its management team but also by its ability to mobilize the right sponsors/board members and experts/partners.

In particular, many ventures fail to invest sufficient time and attention in the recruitment and management of an effective and experienced board of directors or steering committee, and to leverage the resources, legitimacy and strategic vision they can provide. Such ventures often end up moving quickly but in the wrong direction.

Conversely, a venture team supported by many strategic thinkers but lacking effective doers might compile a lot of analysis but will often go nowhere.

Teenage Ventures and Corporate Stepmothers

The second key challenge regarding the governance of an innovative venture is the level of autonomy of its management team vis-à-vis its main stakeholders, particularly its shareholders (for a start-up) and its corporate parent (for a corporate venture).

During the successive dot.com, digitization and artificial intelligence booms, millions of dollars of investor and corporate money were wasted in pointless ventures, because these investors and corporations did not understand what they were investing in but felt they could not afford to "miss the boat".

Innovative start-ups with weak or passive shareholders risk being overwhelmed by the ego and personal ambitions of their managers. But conversely, a venture with an excessively dominant shareholder or micro-managing investors will often fail to attract or retain the entrepreneurial talent it needs.

Too often, young entrepreneurs desperate to raise their first round of financing end up with a motley collection of friends and family, business angels and public investment funds, with various and often incompatible (or even irrelevant) skills and expectations. Managing all these "stepmothers" often consumes most of the entrepreneurs' time and attention at the expense of their business. Experienced entrepreneurs, by contrast, know the importance of attracting "smart" money and designing early effective shareholder agreements.

Maintaining the right level of autonomy is even more important for a corporate venture, for which the access to the resources of its corporate parent is both a challenge and a source of competitive advantage. Designing a successful business model for a corporate venture therefore means designing a governance structure that provides the right level of autonomy, anywhere from a fully integrated corporate project to, at the other extreme, an autonomous spin-off. The right level of autonomy will be a function on one hand of the parent corporation's level of entrepreneurial culture and on the other hand of the venture's strategic importance and operational proximity.

Corporate ventures with badly designed governance structures and processes end up with the worst of both worlds: they have to cope at the same time with the limited resources of an independent start-up and with the bureaucracy of a corporate project. On the other hand, the best corporate ventures can rely on an innovation team combining an entrepreneurial venture team with the best corporate experts.

Let us stress that the autonomy of the corporate venture is both a formal—Who decides what?—and a symbolic—Are they one of us?—issue. If the venture is perceived as an alien project by the corporate staff, it will too often struggle to get their support.

Designing specific packages, such as performance-driven incentives, dedicated facilities or priority access to top management, can be a strong source of motivation for the innovative venture team. But it can also be a hidden source of jealousy and hostility from the corporate staff they will have to rely on, those who feel that they run the "real" business and earn the money that the venture is spending.

So What?

Designing a successful business model also means thinking about the right governance structure, providing on the one hand strategic guidance and operational support and on the other hand the right level of autonomy. Most innovative ventures fail because of bad governance, not bad technology.

5.3.3 Financial Resources: How Much?

Having identified a competitive market opportunity and the entrepreneurial talent and partners to capture it, the last missing piece of the business model puzzle relates to the potential profits and necessary investments.

The majority of new ventures are actually started with very limited means. But most high-risk/high-potential innovation opportunities require significant upfront investments, be it investors' funds or corporate resources. And as soon as significant resources must be committed to capture an innovation opportunity, the question of how much is really at stake must be addressed, involving a careful assessment of what funds are needed, for how long and whether they could be raised.

The average start-up is launched with a few thousand euros (or yuan or dollars) of personal savings or credit card borrowing, and is quickly self-sufficient (or dead). But launching a corporate venture that will generate significant value can take more time (five to ten years) and money than most executives or shareholders are ready to commit. The result is potentially attractive opportunities that are actually never launched, or not given enough time to deliver.

Numbers Speak Louder Than Words

Assessing how much is at stake, in particular whether the value that could be created and captured is greater than the opportunity cost of the resources that will need to be mobilized, is a necessary step each time significant investments are involved. All firms have limited resources and even not-for-profit organizations should consider whether a given opportunity could represent the best use of them.

Even large foundations and super-rich entrepreneurs pursuing their pet projects have limited means and must make tough decisions regarding whether investing their time and resources in an innovation opportunity could represent a good use of resources. And not deciding is actually also a decision, and often not the best one.

While such judgments are often eminently subjective and loaded with "gut feeling" assumptions, simple "back-of-the-envelope" calculations can provide convincing arguments regarding whether a potential business model should or should not be considered. It can also provide some indications of the necessary conditions—"what you need to believe"—for a given business model to be successful.

The promises of many disruptive start-ups often rely on heroic implicit assumptions regarding their future market share and profit margins. Too often, investors forget that even in the new "digital world", the laws of physics still apply and there (still) cannot be more customers than there are human beings.

Garbage In, Financials Out

To borrow Albert Einstein's words, a financial model should be "as simple as possible, but not simpler". This means that the key challenge when evaluating the financials of a business model is not using fancy mathematics or never-ending Excel spreadsheets. The key challenge is making the right assumptions regarding (1) the potential market and market share, (2) the readiness to pay of potential customers and (3) the main required fixed and variable resources.

These must be assessed based on what is known or believed regarding the potential opportunity ("why/why us"), the potential value proposition and value chain ("what") and the potential organization and partners ("who"). If these have been carefully assessed, turning them into a financial number only requires primary-school mathematics.

Too often, innovators or entrepreneurs simply fill financial templates that they do not understand or believe in. In the worst cases, they end up compensating quality with quantity or camouflaging the weaknesses of their business model behind unrealistic numbers carefully hidden in the depths of their financial models.

Paying for Your Cake and Eating It

While it is very easy to generate pages and pages of complex but completely hypothetical numbers, there are three fundamental metrics that should be considered one way or another when considering the financials of a potential business model: the "size of the cake", the "price of the cake" and the "cooking time". Failing to understand these or assessing them based on the wrong assumptions means failing to correctly identify the financial implications of a business opportunity.

First, the "size of the cake" relates to the potential value at stake. Whether measured in terms of marginal cash flows, profits, sales or number of users, it must provide an indication of the potential size of the opportunity which is pursued. Such assessment can allow investors or executives to quickly eliminate opportunities that have insufficient potential given their expectations. It can also help entrepreneurs and innovators to frame or perform a reality check on their ambitions.

The huge size (up to $10tr) of the "mobility and transportation" cake is what has motivated investors and corporations to invest billions in very risky but potentially very profitable new business models and mobility solutions. "Sustainable energy" and "global health" are other examples of "huge cakes" pursued by innovative ventures. But from Segway to Groupon or Second Life, the innovation graveyard is also full of supposedly huge opportunities that never materialized.

Second, the "price of the cake" relates to an assessment of the total amount of resources that will have to be committed over time, until the opportunity can have a chance to be captured. It is important in relative terms (with respect to the "size" of the cake) but also in absolute terms, as the ability of the project to raise or commit upfront the necessary funds is a condition for the business model to succeed. Many potentially attractive opportunities have failed in the "Valley of death" because the "price of the cake" had been underestimated and as a consequence insufficient resources were raised or budgeted.

One of the reasons behind the successes of Warren Buffet is his access to vast funds, allowing his businesses to make commitments and take risks others cannot afford.

Finally, the "cooking time" relates to the time needed for the business model to "pay back" the resources that were committed. It allows for assessing when the project could start to create value and whether the pace of the project is in line with the patience and expectations of its main stakeholders. Many innovation opportunities end up being rejected because their "time-to-value" is too long and can only be supported by public funding or long-term investors.

Aeronautics and healthcare are examples of industries where many innovation opportunities are accessible only to organizations with long time horizons and deep pockets. Conversely, redesigning a business model in order to decrease the payback time and/or the upfront investment provides an opportunity for innovation managers to increase its attractiveness, particularly in industries with short time horizons.

So What?

Whether as a not-for-profit organization or a financial investor designing a business model based on an innovation opportunity, financial implications should always be addressed upfront. This does not require fancy mathematics or never-ending Excel spreadsheets but rather a careful assessment of the potential size, cost and pace of the innovation opportunity, based on the right assumptions regarding the market and competition ("why/why us"), value proposition ("what") and organization ("who").

5.4 Valuating Innovative Business Models: Quantifying the Unquantifiable

Having identified a potential competitive market opportunity based on an innovation and having assessed the implications in terms of human and financial resources, the toughest determination remains whether the opportunity is actually worth it, based on what is known, what is believed, and the remaining uncertainties regarding the opportunity and its alternatives.

This means on the one hand finding a way to aggregate all the information available into a single number—the "value" of the business model—(Fig. 5.5) and on the other hand integrating the key known uncertainties surrounding that number.

Lastly, it means making sense of and building a consensus regarding what those numbers mean and don't mean in terms of a final go/no-go decision.

Key Insights

i. Valuation is a decision-making process aimed at *making smart bets* on the "least bad" way to "put a number" on a business model, taking into account what is known, what is believed and the remaining ambiguities and uncertainties. Numbers are a necessary evil.

ii. The *net present value (NPV)* method is a very powerful but also very dangerous technique for valuing a business model based on its expected riskiness and potential cash flows. It can in particular be effectively used to compare multiple business models under similar cost of capital and terminal value assumptions.

iii. Sensitivity analysis and scenario planning approaches can be used to *integrate known risks and potential uncertainties* and to reduce the scope of managerial ignorance, provided they are based on challenging business conversations regarding "what you need to believe" rather than black box models and/or blind number-crunching.

Fig. 5.5 Valuating innovative business models

5.4.1 The Valuation Decision-Making Process: Making Smart Bets

Making a decision based on the valuation of a potential business model requires on the one hand choosing specific metrics to value the business and on the other hand understanding their intrinsic limitations and the implications of the underlying uncertainties and remaining ambiguities.

The best poker players know how to consistently make the most of what they have in hand (what they perceive and what they know) but also when to fold given the remaining uncertainties and potential outcomes (what they know they don't know).

The Ingredients, the Recipe and the Chef's Touch

Aggregating the available data, information, knowledge, assumptions and beliefs regarding a potential business model—the "ingredients"—into a single value number—the "valuation"—requires a "recipe", a method to convert these inputs into a single metric.

While the NPV of the future cash flows is probably the most popular metric, other choices such as return on investment (ROI), internal rate of return (IRR), payback time or various multiples are also often explicitly or implicitly used by investors and executives in assessing innovative business models.

Early stage investors often use metrics based on multiples of sales or profits (if any) when assessing innovative ventures, comparing them with the valuations of similar deals. More exotic measures such as value per user, momentum or option pricing can also sometimes be used. In the worst cases, the investors will use what-ever metrics actually confirm their "gut feeling", or convince themselves that "this time it is different".

Making a decision based on such metrics also requires defining an explicit or implicit threshold—the "chef's touch"—regarding the minimum accept-able hurdle for them. Explicit thresholds can relate to maximum time to mar-ket, minimum sales volumes or minimum levels of profitability. Implicit thresholds are used, for example, when defining the risk premium needed to compute the discount rate in an NPV calculation.

The threshold can be different for different types of innovations. Typical thresh-olds used by industrial firms include an ROI of 12–15%, a payback time of less than three or less than five years, or a discount rate of 10–12%. As another exam-ple, venture capitalists typically use a payback (exit) time of five to seven years and a discount rate closer to 25–30% for early stage ventures.

An important consequence of the use of metrics and thresholds is that smart innovation managers who design innovative business models should know, take into account and master the metrics and thresholds that lie in store for them. A business model which delivers a good payback time or ROI might not convince an investor focused on NPV, sales growth or earnings multiples.

Know You Don't Know

Making a decision based on a single metric has the benefit of simplicity. It also facilitates the comparison of different business models or different versions of the same business model.

But such simplification also comes with a significant cost, which must not be ignored. Not only will the choice of metric influence the valuation's out-come, but, even more important, the resulting value is always loaded with assumptions and uncertainties. Using a valuation number without under-standing those is often even worse than advancing in the dark.

History is full of cases where supposedly "smart" investors and "efficient" markets used valuations based on completely unrealistic or heroic assumptions, be it regard-ing Dutch tulips, exotic technologies, Silicon Valley unicorns or "Initial Coin Offerings".

The "known unknowns" that innovation managers have to deal with include various types of organizational, technical and commercial risks, for which both a probability distribution and a cost can be estimated. But they also include intrinsic uncertainties, situations when the past does not provide any reliable guide to future events, and where therefore it is not possible to assign a range or probability distribution. Finally, the complexity of innovations means that managers also often have to face ambiguity, information for which multiple interpretations are possible.

Risks include variations in market size, required investment or time to market, which can all affect the expected value of a business model. Uncertainties include the consequences of disruptive innovations (e.g. the internet), of catastrophic events (e.g. Fukushima) or of macroeconomic evolutions (e.g. Brexit). Ambiguous situations include unexpectedly low profitability or slow sales growth, which can have multiple, complex and interdependent causes.

Agreeing on the Least Bad Bet

Valuation techniques are powerful decision-support tools. And like most powerful tools they should be used with caution. Numbers are necessary evils, full of uncertainties and assumptions, but are often more relevant—and convincing—than simple gut feelings.

"What gets measured gets done" is a well-known business mantra. But to borrow words attributed to Einstein, *"Not everything that can be counted counts, and not everything that counts can be counted".*

Valuation techniques are particularly useful on the one hand when comparing business models using similar assumptions and on the other hand when testing multiple assumptions in order to understand the key value drivers of a business model. These techniques can help identify sources of disagreement and facilitate consensus.

Valuation is a business analysis, decision-support and consensus-building technique, not a substitute to business acumen.

So What?

Valuation techniques provide a powerful way to aggregate in a single number—the "value"—all the available inputs regarding a potential business model. They can effectively support decision-making and consensus-building regarding an innovation opportunity, if used with caution, particularly regarding the way beliefs, assumptions, ambiguities and uncertainties are dealt with. Innovation managers should therefore adapt their valuation approach to the context and objectives. They must in particular avoid drowning in numbers and always "know they don't know". Valuation is more art than science.

5.4.2 Carefully Using NPV to Value Business Models

The NPV, also called the discounted cash flow (DCF) method, is probably the most widely known and commonly used method to financially value a business in general and an innovative business model in particular. However, its popularity hides major weaknesses, which have long been known in theory but are too often ignored in practice. It is a very useful method but should be handled with care.

Mother of All Valuation Methods

The key strength of the NPV method is that it integrates a wide set of parameters that most other methods ignore. First, it takes into account "all" the net cash flows, and not only those needed to pay back the initial investments. Second, it takes into account the time value of money, that is, the fact that one euro today is worth more than the same euro tomorrow. Finally, it takes into account the opportunity cost of taking risks, that is, the fact that one safe euro is worth more than one risky euro.

First, take all the future cash flows you can compute and pick a cost of capital to discount each of them. Second, discount each cash flow in line with its timing and pick a way to deal with all the remaining cash flows. Add all the things up, stir a bit, and serve!

Handle with Care

The completeness of the NPV method is one of its key strengths. But this completeness comes at the cost of additional assumptions whose potential implications should not be ignored when valuing innovative business models. Indeed, on top of having to assess the potential future net cash flows linked with a business model, the NPV method also requires further assumptions regarding the cost of capital, the final value and the risk profile. And these further assumptions can have a huge impact on the result of the valuation.

Too often, the assumptions regarding the cost of capital and the final value are taken for granted, ignored or hidden in the footnotes or appendices of the valuation report. In most cases, the risk profile assumptions are not even considered.

The first assumption relates to the cost of capital needed to discount the future cash flows. While all finance students learn some formulas for calculating cost of capital, such as the weighted average cost of capital (WACC), these calculations require, one way or another, making an assumption regarding the underlying riskiness of the business model. And when valuing an innovation opportunity, assessing its riskiness can be very challenging.

Many organizations skip this challenge by actually fixing a cost of capital that must be used a priori *for valuating all their innovation projects. Others explicitly derive this cost from the risk premium associated with similar ventures (the "beta"), but this also requires an assumption regarding which ventures are considered "similar".*

The second assumption relates to the "final value" of the opportunity. As cash flows cannot be computed and discounted forever, an assumption must be made regarding the value of all the remaining cash flows, those occurring after the last period of evaluation of the opportunity. This value can be assumed to be zero or computed using standard actuarial methods. The key issue here is that in many cases the hypothetical value both is highly uncertain and can represent a significant share of the total NPV which is computed.

When considering innovation projects with a clearly limited timespan (e.g. a fashion item or equipment), the final value can be computed as zero or even negative (if there are end-of-life recovery costs). However, when developing a new business, assuming that the "end-of-life" value is zero or negligible can be a very biased assumption.

The third assumption relates to the risk profile of the innovation opportunity as it unfolds over time. One key assumption of the NPV method is that the riskiness of a project can be modeled through a discount rate. This implies, among other things, that the "risk penalty" of an innovation increases exponentially over time at a constant rate. While this can be a reasonable assumption when borrowing money in the short term and at a fixed rate, one should not assume that risks always affect value in such a way. Using variable discount rates can be an option, but this adds complexity and requires further assumptions.

The time bias of NPV methods can raise tricky questions when dealing with very long-term projects and "future generations", for example, related to nuclear waste or global warming. An innovation that could have enriched a Roman emperor but would then destroy the whole earth today could actually have had a positive NPV.

The Value of Flexibility

A final weakness of NPV methods is that they often do not take into account how much a business model—and therefore its future cash flows—can be adjusted over time. They fail therefore to put a value on the flexibility of a given business model, such as the future possibility (or not) of abandoning or deferring a project, expanding or shrinking it, closing and restarting its development or switching some inputs or outputs. Option pricing approaches (Adner and Levinthal 2004) based on stochastic models have been developed and can be used to value such flexibility but, again, only at the expense of increased complexity and further assumptions.

An initial investment in R&D can be considered the purchase of an option, bestowing the right—but not imposing the obligation—to invest in a future business development. If the innovation is successful the business development costs can be considered the price to exercise the option that was purchased, and the future profitability of the business is the underlying asset.

So What?

NPV or discounted cash-flow approaches allow for aggregating into a single number the expected cash flows and perceived riskiness of a business opportunity. They rely, however, on major assumptions that cannot be taken for granted when dealing with innovation opportunities. Innovation managers should consider NPV approaches as useful but fragile decision-support tools, not infallible oracles.

5.4.3 Integrating Known Risks and Potential Uncertainties

One way to address the intrinsic limitations of valuation techniques when dealing with innovation opportunities is to complement the calculated "value" number with further information regarding its relative robustness. This allows innovation managers not only to assess the "most likely" financial value of an innovation opportunity but also to assess the "strings attached", that is, the level of confidence they can put in that valuation. This can be done by integrating known risks, for example, through sensitivity analysis, and by addressing potential uncertainties, for example, through scenario analysis.

Sense and Sensitivities

When designing an innovative business model, several key parameters and assumptions underlying the innovation opportunity are often identified as a priori susceptible to significant variations. In some cases, past experience or simulations can be used to try to assign a range and/or probability distribution to the potential variations. It is then possible to change the value of the "critical" parameters and to analyze the impact of such risks on the expected value of the opportunity. This can be done by changing one parameter at a time, by adopting pessimistic (worst case) and optimistic (best case) assumptions regarding several of them or by running large-scale statistical simulations.

Critical parameters that can significantly affect the valuation of an innovation opportunity include input prices, market share, development time and cost of capital, which can influence its investment cost, market potential, payback time and ultimately its NPV.

While generating new numbers in such a way can be relatively easy, drawing clear managerial implications in terms of business model design and go/no-go decisions often remains challenging. In particular, some critical parameters can be neglected, their variance can be underestimated through overconfidence, or the misunderstanding of their interdependencies can lead to overlooking significant systematic effects.

A sensitivity analysis whose only result is that in the best/worst cases the value of the opportunity is higher/lower is trivial. Playing with numbers in order to generate more numbers is too often meaningless. As another example, in 2011 and 2014 successive fluctuations of the exchange rates of the Swiss Franc severely disrupted financial markets, because these fluctuations had been assessed as completely impossible by the statistical models used by most financial institutions.

Better Analysis, Not Bigger Spreadsheets

While sensitivity analysis can be used to analyze the impact of known risks, they often fail to address the complexity of innovative business models and the uncertainties that the complexity generates. Instead of trying to identify the business implications of changing numbers, it is often more useful to try to analyze the quantitative implications of changing businesses.

In such a scenario-based approach, the first step is to analyze the business model and its context in order to identify possible relevant evolutions, each leading to a specific scenario. Traditional valuation techniques are then used in order to assess the potential value implied by each scenario. A key challenge is of course to generate "what if" scenarios that are both unlikely but relevant—not to be narrow-minded—and representative but not overwhelming—not to be lost.

Business scenarios can be generated by, for example, considering potential evolutions of technologies, markets, competition, regulation or socioeconomic factors. The "business as usual" scenario of not innovating should also be considered an alternative. Examples of relevant but unexpected scenarios include 9/11, the

Enron bankruptcy, Hurricanes Sandy and Katrina, Fukushima, the global financial crisis, the election of Donald Trump and Brexit.

What You Need to Believe

One way to interpret the information thus generated is to identify the necessary assumptions for the business to be profitable: "What you need to believe". Innovation managers can then try to identify early-warning indicators in order to anticipate as much as possible whether the assumptions could be met. Another way to deal with the assumptions is to integrate within the design of the business model some fallback options, in order to be able to cope with alternative outcomes.

Scenario-based approaches can therefore be more holistic and business-driven than sensitivity analyses. They remain, however, challenging given the potentially large number of scenarios to consider and analyze, and the failure of most people to be ready to consider scenarios that are really radical and/or strongly challenge their pre-existing beliefs.

Examples of psychological biases that weaken scenario-based approaches include the "typical things" bias (the future will be like the present), confirmation bias (the future will be like I believe it will be) and herd behavior (we all seem to agree how the future will be).

So What?

When assessing innovative business models, managers should complement traditional valuation techniques with risk assessment and mitigation approaches such as sensitivity and scenario analysis. The challenge, however, is to reduce the scope of their ignorance, not just to fill spreadsheets. Overconfidence or "analysis paralysis" are often worse guides than ignorance.

5.5 Building a Consistent and Balanced Innovation Portfolio

An organization that considers an innovative business model should obviously assess its intrinsic value before deciding whether to pursue it. But that is not enough. The decision whether to pursue an innovation opportunity should also be taken based on the impact the opportunity would have on the portfolio of existing projects, both innovative and non-innovative.

This means considering on the one hand whether the new opportunity is consistent and compatible with the existing portfolio, and on the other hand how and how much it could affect the overall balance and alignment of that portfolio (Fig 5.6).

Key Insights

i. The value of an innovative opportunity—"the egg"—must be assessed, taking into account how and how much it fits with the *portfolio* of innovative and non-innovative projects currently pursued by an organization—"the baskets".

ii. How an innovation opportunity will affect the *consistency* of the corporate portfolio should be assessed, in terms of both the potential critical resources bottlenecks—in particular management's and customers' attention—and the potential technological and organizational synergies.

iii. How much an innovation opportunity will affect the *balance and alignment* of the corporate portfolio should be assessed, both in terms of strategic scope—exploitation versus exploration—and with respect to the time horizon—short, mid and long term.

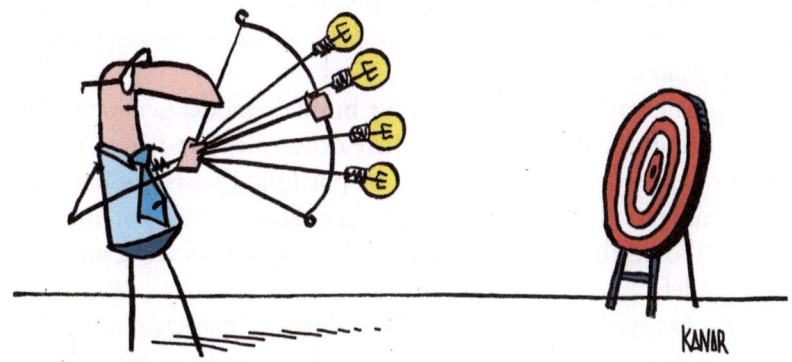

Fig. 5.6 Building a consistent and balanced portfolio

5.5.1 Managing a Portfolio of Opportunities

The performance of an organization will be based not only on the cumulative contributions of all its initiatives, but also on how much those initiatives form a consistent and balanced portfolio (Cooper et al. 2001). This means that the potential contribution of an individual innovation project should be assessed on a stand-alone basis but also taking into account how and how much it affects the corporate portfolio.

Financial investing is also a field where an individual investment, such as the purchase of a stock, must be assessed based not only on its intrinsic expected cash flows but also on how and how much it affects the risk profile of the entire portfolio. The big difference is that in finance diversification is (almost) always a good thing. But in innovation the right level of diversification is a trade-off between risk minimization and focus. Too much diversification means wasted resources, not enough diversification means high risks.

The New Piece and the Puzzle

Innovative business models do not exist in a vacuum. They have to find their place in the corporate agenda and in particular gain access to the corporate resources they need. The most common mistake here is to assume that the needed resources can always be made available "on top of" existing activities, without dedicated management attention. While available slack resources might sometimes be sufficient at an early stage, implementing an innovative business model is real work, requiring real resources.

Too often managers do not recognize the opportunity cost of an innovation opportunity, assuming that the necessary resources can be freed without sacrificing any of the existing activities. The usual result is bottlenecks, never-ending projects and/or burnt-out employees.

The implementation of an innovative business model generates opportunity costs. But it can also generate positive externalities and cross-fertilization opportunities for the organization and its portfolio of existing projects, such as new learning, new customers or new capabilities. These potential positive side-effects should therefore also be considered when assessing an innovative business model.

While the bottom-line impact of the "Solar Impulse" project launched by Solvay was probably negative in the short term, it generated huge positive side-effects in terms of reputation, employee engagement, organizational learning and technological developments. Similarly, Google's "Moonshot" projects have so far contributed negatively to its cash flows but positively to its innovative image and employer brand.

A New Egg in the Corporate Baskets

An organization must develop and maintain a portfolio of innovative and non-innovative projects that is aligned with its strategic objectives. Launching an additional innovation project can affect the portfolio's balance, be it in terms of risk profile, exploitation versus exploration or short- versus long-term investments. Assessing an innovative business model therefore also requires assessing how it will affect the various trade-offs, and how it could potentially un- or rebalance the corporate portfolio.

A corporation with a large number of long-term projects might more favorably consider a short-term innovation opportunity. Conversely, a business with mostly defensive projects and limited growth prospects might prefer launching new high-risk/high-potential disruptive innovation projects. Finally, a firm whose current market and market share are growing fast or that is in the middle of a significant restructuring or capacity ramp-up might temporarily freeze new innovation projects.

So What?

Assessing an innovative business model means also assessing how it will affect the corporate portfolio of activities, in terms of the opportunity costs of its critical resources, the potential side benefits it might generate and how and how much it will affect the portfolio's balance and alignment. A good innovation opportunity is not always the right one.

5.5.2 Building Consistent Innovation Portfolios

Except in the case of start-ups, most organizations that consider an innovative business model will also be managing other activities and projects. And except in the case of fully independent spin-out ventures, the innovative business model they consider will interact and share some resources with their other activities and projects. Assessing an innovative business model therefore means also assessing the impact and value of these interactions, both negative (opportunity costs) and positive (synergies).

Avoiding Corporate Bottlenecks

One of the main reasons a corporation might be well positioned to capture an innovation opportunity is that it enjoys unique and valuable resources, which could be leveraged for such an opportunity. Such resources include, of course, its people, funds and assets, which should not be "sprinkled" over dozens of projects. But they also include the support of its management and access to its customers' pockets, resources that are valuable but scarce and that need to be carefully managed in order to avoid unexpected bottlenecks.

Most organizations have explicit resource allocation and prioritization processes for their people, funds and assets. But many fail to also carefully manage the limited attention span of managers and customers. The result is too often an unmanageable executive agenda, cannibalization and overcrowded product launches. Limited managerial resources have indeed long been known to be one of the main factors limiting firm growth.

Innovation projects will therefore compete with existing activities for critical resources, and this competition should be proactively managed. Assessing an innovative business model therefore means also assessing the opportunity cost and availability of the corporate resources it needs.

When a new project and ongoing activities compete for the same corporate resources, the result will too often be driven by who has the loudest voice (in most cases the existing activities), not by what is best for the organization. And assuming that innovation projects can always rely on idle resources or overtime ("on top of") means that innovation could remain "nice to have" rather than a real corporate priority.

1 + 1 = 3

Capturing an innovation opportunity will generate new cash flows, be it investments, costs or revenues. But it will often also generate new resources and capabilities that could benefit the corporation as a whole, and the potential value of these new resources and capabilities should also be considered when assessing the opportunity. Even a project that is stopped or "fails" from a cash-flow point of view often generates knowledge and/or resources that have value and can be recycled by the corporation.

Large firms often launch long-term exploratory projects, for example, in fields such as artificial intelligence or bioengineering, which are not expected to directly generate positive cash flows but will allow the firms to learn and develop potentially valuable capabilities. Similarly, the valuation of disruptive start-ups often reflects not only their expected future cash flows but also the upside potential of the capabilities they build, for example, in terms of big data. Assessing such projects only on a cash-flow basis is often short-sighted.

Here again the challenge is not only to assess the potential synergies but also to manage them proactively. This is particularly the case for projects running on different time-scales, or for projects where third parties or separate corporate entities are involved. Indeed, in both cases knowledge and resource sharing and transfer will not "naturally" cross organizational boundaries.

One way to manage such positive externalities is through the definition of explicit corporate or industry technology roadmaps, where the competencies developed in one project are managed to support the development of technology platforms needed for a second project, which in turn is used to develop new products, processes or functionalities in other projects. The real value lies in the whole roadmap, not only in the potential cash flows of individual projects.

So What?

Managers who assess the value of an innovative business model should consider not only its intrinsic cash flows but also its impact on other projects and activities, both in terms of the (negative) opportunity cost of resource bottlenecks and the (positive) externalities of project synergies and cross-fertilization. An innovation portfolio should be worth more than the sum of its project cash flows.

5.5.3 Staying the Course: Balanced and Aligned Portfolios

Managing a portfolio of innovative and non-innovative activities requires striking a careful balance between various and often conflicting objectives, such as focus versus diversification and short-term profitability versus long-term growth. Assessing an innovative business model therefore also means considering how and how much it could disrupt or improve this balance, in other words whether it will lead the corporation's positioning and capabilities closer to or further away from its strategic priorities.

Mining Versus Hunting

A balanced portfolio will combine on the one hand opportunities identified within the current businesses—focus—and on the other hand opportunities to be created outside the current scope of the organization—diversification.

In the first case, the focus is on "filling the blanks", by improving existing products, technologies and processes in order to more efficiently serve existing markets. In the second case, the focus is more on "breakthrough" opportunities, new business models involving products, technologies, processes and/or markets that are new to the organization in particular or even to the world in general. The right balance for an organization regarding this trade-off is a strategic choice, driven by its objectives, resources and environment.

Innovative digital technologies can be used by incumbent firms both to more efficiently run their existing businesses and to build completely new businesses. Similarly, sustainability issues can affect both existing activities (e.g. in terms of carbon footprint) and create opportunities to launch new ones (e.g. new energy management services).

Assessing an innovative business model therefore requires identifying where it sits in terms of existing/new business opportunities in the corporate portfolio, particularly regarding its technological and commercial feasibility. It requires then assessing whether it will tilt the overall portfolio in the right direction, either in terms of target product/market positioning or in terms of capability development.

Industrial firms often overemphasize the value of R&D-based radically new products, while they should probably invest more resources in innovation projects aimed at improving the efficiency and sustainability of their existing organization and activities. Conversely, many financial institutions have focused too much on their traditional activities and failed to capture new digital business model opportunities.

Today Versus Tomorrow

A balanced corporate portfolio will include projects and activities whose costs and benefits are spread over time, in line with the organization's resources, environment and objectives. Assessing an innovative business model will therefore also mean assessing how it could positively or negatively affect this balance.

Pharmaceutical and aeronautics companies are used to considering innovation opportunities with expected cash flows spread over years or decades. Consumer electronics or entertainment firms are used to projects whose lifespan covers months or quarters. The same innovation could therefore obviously fit one but not the other.

There are in particular three time horizons that have been identified as relevant when assessing an innovative business model, each linked with different performance indicators.

First, in the short term, an innovative business model could strengthen or weaken the profitability of existing activities—the core business.

Second, in the mid-term, an innovative business model could contribute to the development of opportunities already identified and/or launched by the organization—the new businesses.

Third, in the long term, developing an innovative business model could create opportunities to develop future but not yet identified businesses—future options.

In the first case, the innovative business model should be assessed mainly in terms of its impact on short-term profitability; in the second case, of mid-term sales growth; and in the third case, of new long-term capabilities. Again, the exact timespan—in months or decades—to which the time horizons refer will be a function of the firm, industry and environment.

A typical start-up will have to balance long-term growth opportunities (e.g. in biotech or big data) with short-term liquidity needs (e.g. by offering consulting services). Similarly, an incumbent corporation must innovate to defend its core businesses but also to create new ones. Again, the same innovation could therefore obviously fit one but not the other.

So What?

An innovative business model must be assessed not only in terms of its operational costs and benefits to the organization and of its activities, but also in terms of its strategic fit with the corporate portfolio, both in terms of risk/potential profile and time horizon. A good innovation opportunity might not be the one the corporation needs.

5.6   Synthesis

Develop a Balanced Portfolio of Business Models: Key Insights

5.1. Business model design: asking the right questions

i. Designing or improving a *successful business model*, business planning, means on the one hand building and validating a credible story regarding how specific resources could be (better) mobilized to solve a specific problem and on the other hand selling that story to the relevant internal and external stakeholders.
ii. A business model will be convincing if it addresses in a consistent way four *key questions*: (1) Why is there a problem and why are we well positioned to solve it? (2) What exactly could be sold to whom and how? (3) Who needs to be mobilized? (4) How much is at stake?
iii. Finding new ways to address the why, what, who and how much key questions around the same innovation opportunity can allow managers to *design innovative business models*.

iv. Successful entrepreneurs and investors *do not plan to fail*. They prioritize the "why?" and "who?" key questions when assessing an innovative business opportunity. They know that the technology specifications ("what?") and financial spreadsheets ("how much?") will change and will need a lot of time and further effort to be fixed.

5.2. Designing competitive business models: Why and what?

i. The first (obvious but still too often forgotten) strategic challenge of a successful business model is to *address somebody's problem*. It must demonstrate that potentially enough people out there have a problem and that the problem is painful enough for them to be ready to both adopt and pay for a new solution.

ii. The second strategic challenge of a successful business model is to be *better positioned than others*. The innovators and their organization must demonstrate that they could address a relevant problem or meet a need better than the available alternatives, based on their scope as well as their scale, unique assets and/or agility.

iii. On the other hand, the first operational challenge of a successful business model is to conceive an initial *value proposition*, which some people will actually be ready and able to find, adopt and pay for. What will be on the first invoice?

iv. The second—and too often forgotten—operational challenge of a successful business model is to set up, integrate and scale up over time a *competitive value chain*, including the right design, operations, client management and support activities.

5.3. Mobilizing the right resources: Who and how much?

i. Probably the most important but least understood aspect of a successful innovative business model is the identification and mobilization of the required *entrepreneurial talent and expertise*. Too often firms decide first to launch a project and then try to staff and link it with whoever is available, that is, in many cases not the right people or partners.

ii. The hidden secret behind the failures of many innovative ventures is not bad technologies but bad *governance*, with the wrong people taking the wrong decisions or failing to take any decisions at all.

iii. The easiest way to waste an innovation opportunity is to underestimate the *financial resources* required over time to support the launch and growth of a sustainable business model. Great innovations are never overnight successes.

5.4. Valuing innovative business models: quantifying the unquantifiable

i. Valuation is a decision-making process aimed at *making smart bets* on the "least bad" way to "put a number" on a business model, taking into account what is known, what is believed and the remaining ambiguities and uncertainties. Numbers are a necessary evil.

   ii. The *NPV* method is a very powerful but also very dangerous technique for valuing a business model based on its expected riskiness and potential cash flows. It can in particular be effectively used to compare multiple business models under similar cost of capital and terminal value assumptions.

  iii. Sensitivity analysis and scenario planning approaches can be used to *integrate known risks and potential uncertainties* and to reduce the scope of managerial ignorance, provided they are based on challenging business conversations regarding "what you need to believe" rather than black box models and/or blind number-crunching.

5.5. Building a consistent and balanced innovation portfolio

   i. The value of an innovative opportunity—"the egg"—must be assessed, taking into account how and how much it fits with the *portfolio* of innovative and non-innovative projects currently pursued by an organization—"the baskets".

  ii. How an innovation opportunity will affect the *consistency* of the corporate portfolio should be assessed, in terms of both the potential critical resources bottlenecks—in particular management's and customers' attention—and the potential technological and organizational synergies.

  iii. How much an innovation opportunity will affect the *balance and alignment* of the corporate portfolio should be assessed, both in terms of strategic scope—exploitation versus exploration–and with respect to the time horizon—short, mid and long term.

Bibliography[1]

Adner, R., & Levinthal, D. A. (2004). What is not a real option: Considering boundaries for the application of real options to business strategy. *Academy of Management Review, 29*(1), 74–85.

Cooper, R., Edgett, S., & Kleinschmidt, E. (2001). Portfolio management for new product development: Results of an industry practices study. *R&D Management, 31*(4), 361–380.

Delmar, F., & Shane, S. (2003). Does business planning facilitate the development of new ventures? *Strategic Management Journal, 24*(12), 1165–1185.

Dierickx, I., & Cool, K. (1989). Asset stock accumulation and sustainability of competitive advantage. *Management Science, 35*(12), 1504–1511.

Foss, N. J., & Saebi, T. (2017). Fifteen years of research on business model innovation: How far have we come, and where should we go? *Journal of Management, 43*(1), 200–227.

---

[1] An extended bibliography is available at www.NavigatingInnovation.org

Haynie, J. M., Shepherd, D. A., & McMullen, J. S. (2009). An opportunity for me? The role of resources in opportunity evaluation decisions. *Journal of Management Studies, 46*(3), 337–361.

Teece, D. J. (2010). Business models, business strategy and innovation. *Long Range Planning, 43*(2), 172–194.

6

Nimble Execution: Fail Fast and Win Big

An innovation opportunity is worthless if it remains only an idea, a concept or a business plan. Successfully managing innovation therefore means developing the capabilities to effectively capture the innovation opportunities that the organization identifies as attractive. But innovation opportunities cannot be managed like "normal" corporate projects, where uncertainties are minimized and failure is the exception. Nor can they be managed "on top of" existing activities and budgets, without the right processes and resources.

The fifth and last innovation management challenge is therefore to effectively capture innovation opportunities through dedicated project management and decision-making approaches, balancing effective execution with timely learning and flexibility, and mobilizing the right funding sources (Fig. 6.1).

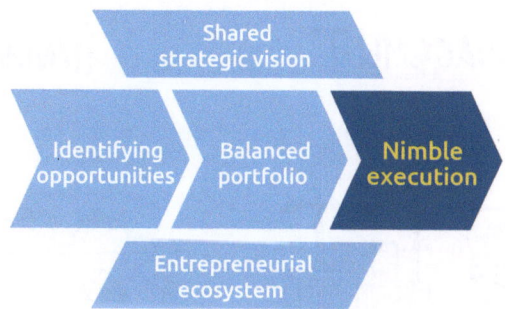

6.1 Nimble execution: learn cheaply and adapt quickly
6.2 Lean development: more brain, less storming
6.3 Smart money: funding innovation projects
6.4 Synthesis

Fig. 6.1 Nimble execution: fail fast and win big

© The Author(s) 2018
B. Gailly, *Navigating Innovation*, https://doi.org/10.1007/978-3-319-77191-5_6

6.1   Nimble Execution: Learn Cheaply and Adapt Quickly

Managers who implement innovation projects must move away from a fully deductive management approach, where knowledge is a given, key performance indicators are fixed and people are "just" FTEs (Full Time Equivalents) that must be allocated and monitored. Innovation managers must adapt their decision-making and project management approaches to a world they know they don't know, where objectives and environments are shaped rather than given and the challenge is to be convincing, not just convinced (Fig. 6.2).

Key Insights

i. Management is (maybe) a science, but *innovation is an art*. Traditional corporate decision-making and project management approaches are therefore ill-suited to the ambiguity, high failure rate, pace and multifunctional aspects of innovation.

ii. *Innovators must play poker* rather than chess. Traditional corporate decision-making approaches ("chess") rely mainly on analyzing facts and minimizing failures—"thinking first". But managing innovative organizations and capturing sizeable innovation opportunities ("poker") must also rely on experimentation and proactive learning, combined with ruthless prioritization—"doing first".

iii. While traditional project management approaches rely mainly on set targets and task allocations, managing innovative organizations and capturing innovation opportunities also implies *planning for changes in the plan* and focusing on embedded flexibility and cross-functional mobilization.

Fig. 6.2   The science of management and the art of innovation

6.1.1 Management Is (Maybe) a Science, but Innovation Is an Art

The dominant view of management as a science is built on the assumption that decisions should be based on known facts and predefined objectives. From such a deterministic and linear perspective, environment and objectives are given, key performance indicators are selected, options are identified and assessed, valuations are consolidated and the best option is then chosen, implemented and measured. The keywords here are alignment, standardization, optimization, systematization, planning and control.

This is the "hard" world of engineers, financiers and economists, for whom the firm is essentially a production function or a "machine" whose efficiency must be maximized and whose scale must be optimized. In this world, innovations are either good or bad based on their intrinsic characteristics, and good innovations always win.

Many managers still consciously or unconsciously live in a Frederick Taylor or Henry Ford world, where they see their role mainly as (1) analyzing facts, (2) defining targets, budgets and plans based on those facts and (3) detecting and correcting deviations from the plan. This is the world of "what gets measured gets done".

From Just in Time to Just in Case

While a "scientific approach" to management can generate tremendous benefits in terms of efficiency, it is less and less in tune with a world where innovation is the rule, not the exception. In an innovation world, the challenge is to build organizations that are as nimble as change itself. The objective is no longer to be right but rather just to be less wrong and adjust more quickly than your competitors.

Nobody knows whether and when we will all have electric and/or self-driving cars, bioengineering or humanoids with true artificial intelligence. But the winners will be the firms that made the least bad bets regarding the challenges and built the capabilities to learn and quickly adjust along the way.

If You Have No Good Maps, Make Sure You Have a Good Backpack

Innovation managers must cope with the difficulties of "normal" projects, combined with the peculiarities of innovation projects. The peculiarities include high levels of ambiguity and uncertainty, high failure rates, resources and time to market bottlenecks and complex interdependencies between tasks, people and functions.

This is the "soft" world of social ecosystems, where firms are human institutions, people are subjective and adaptable beings and environments and objectives evolve. The keywords here are flexibility, beliefs, opportunism, experimentation, entrepreneurship and engagement.

So What?

Innovation management is neither traditional management nor just about luck and randomness. Managing innovation well matters. But managing well in an innovation-intensive world means adjusting the way decisions are made and projects are managed. It means focusing on learning (play poker, not chess), prioritization (fail quickly), flexibility (plan for changes in the plan), and mobilization (proactively identify and enlist stakeholders).

6.1.2 Play Poker, Not Chess: Learning and Folding

Probably one of the biggest challenges for managers dealing with innovation projects is to recognize and deal with the scope of their ignorance. This means recognizing that when innovating learning is a critical issue, not a side effect, and that failure is the norm, not the exception.

Thomas Edison and, more recently, James Dyson are two innovators who are known to have gone through hundreds or even thousands of trials and errors before successfully developing innovations. But to them the "errors" were not failures; they were experiments from which they learned. The key was to be able to quickly discard failed prototypes and move on to the next attempt, and focus on convincing everybody else when a good solution had been found.

Know You Don't Know What You Don't Know

Scientists have long integrated in their models the notions of noise and uncertainties. They know their answers are not "exact" but should rather be considered the most likely outcomes within limited confidence intervals. These "known unknowns" can be dealt with using past experience and stochastic modeling.

In quantum physics, we cannot simultaneously know where a particle is and what its momentum is, but we can still assign known probability distributions to those measures. As a consequence, we can make calculations and infer quite accurate results based on those probabilities. Even with chaotic models such as weather or gravitational systems, we know what we do not know and we can try to assess outcomes with a reasonable level of confidence.

But innovation is about going where no one has gone before, where there is no or limited experience one can rely on. Innovation managers must therefore cope with their "tacit ignorance" and recognize that they don't even know what they don't know. In such "Knightian" uncertainty, both the alternatives and their probabilities are indeterminable (Wiltbank et al. 2006).

Many large multinationals had not included "9/11" and, more recently, "Brexit" or the election of Donald Trump in their strategic planning. Similarly, many airlines and European businesses thought Eyjafjallajökull was an Icelandic delicacy, not a volcano that could disrupt air travel during several weeks, as it did in 2010.

This means that innovation managers must maximize learning to "find out what they did not know they did not know" and find ways to reduce the costs of the resulting unpredictability and high failure rates.

Designing Cheap, Smart Experiments

The implication for innovation managers is not that they should take risks for the sake of it but rather find smart ways to learn and cope with the risks. This means that they must be ready in some cases to "do first" rather than "think first" (Mintzberg and Waters 1985). It also means that they should find ways to maximize the speed and minimize the cost of such prototyping and learning, by proactively designing and running cheap and smart experiments.

Finally, it means that an innovation opportunity with the highest potential but for which no ways to learn cheaply can be found might not always be the best opportunity to pursue.

Design thinking and "MVPs" (minimum viable products) are examples of approaches aimed at systematically learning in a cheap and smart way. They are particularly popular in internet businesses, where running such experiments by tweaking multiple versions of the same websites and observing user behaviors is particularly simple and easy.

The 5/95 Rule: Not 80/20

The consequence of this experimentation and learning process is that "failure", that is, project termination without meeting initial expectations, is the norm, not the exception. A typical innovation portfolio will include a few winners and many "failed" projects. Traditional managers are expected to achieve most of their objectives; in innovation they should be expected to stop most of their projects.

Even in Silicon Valley, less than one in six million new business ideas leads to a successful IPO. A typical hi-tech firm or investor will convert less than 1% of its deal flow of bright ideas into sizeable businesses or investments. Their challenge is to pick and boost innovative winners, not to salvage laggards.

The challenge for innovation managers is therefore to be able to stop projects and liberate resources from yesterday's priorities. But stopping a project that executives had approved and for which resources have already been sunk is one of the most difficult things for many managers to do. Stopping a project is still seen as lacking commitment and/or recognizing that a mistake was made. But rather than trying to save losing projects, innovation managers must focus on turning winning projects into winning businesses.

Less than 10% of investments made by technology investors become huge successes, but those few successes account for most of the industry's profitability. Like experienced gold diggers, innovation managers must frantically dig deeper where there seems to be some gold and at the same time quickly stop digging where they do not find any.

Why We Still Get Things Wrong: Decision Biases

Knowing that failure is the norm is one thing, being able to behave accordingly is another. Our human brains are still wired in ways that lead most of us to consistently take the wrong decisions when faced with ambiguous, complex, uncertain and/or new situations such as new innovation opportunities. Subjective cognitive biases are now well documented, but escaping them when making executive decisions regarding innovation opportunities remains a challenge.

Cognitive biases identified by recent Nobel Prize winners include confirmation or overconfidence bias (denial of counterintuitive evidence), sunk cost fallacy (factoring unrecoverable costs), escalation (overinvesting through never-ending small steps) and anchoring or framing (inability to question initial estimate or dominant paradigm). Examples of innovation opportunities initially dismissed by experienced investors include Amazon, Apple, Google, PayPal and Facebook.

So What?

A traditional manager is expected to use what he or she knows, to first think hard and then act and control resources in order to reach most of his or her objectives. An innovation manager should be expected to often first act in order to learn in smart ways, and then to adjust his or her objectives and resources in order to stop most projects and focus on a few winning ones. The key is not "to take more risks" but rather to learn cheaply and adapt quickly.

6.1.3 Plan for Changes in the Plan: Flexibility and Mobilization

The implicit assumption of traditional project management approaches is that both knowledge and objectives are essentially given. As a consequence, the focus of a project manager is first to plan, then to act, that is, to implement and control. In particular, the various people involved in a traditional project will be informed when their contribution is expected, their involvement and deliverables will be monitored and corrective actions will be taken if needed.

But when dealing with innovation opportunities, such a deterministic and sequential approach is not effective. When dealing with innovation opportunities, managers need to plan for changes in the plan and proactively engage the key stakeholders early on.

Do, Check, Plan

As innovation opportunities unfold, prediction errors and unknown unknowns will emerge, perceptions and beliefs will drift and new threats and opportunities will appear as markets, industries and environments evolve. Innovation opportunities mutate or even disappear. A plan that was the best one yesterday might not be the best one—or even a good one—today.

Turning an innovation idea into a sizeable and profitable business will typically take five to ten years. But the world we knew five years ago is completely different from the world we know today, be it in terms of geopolitics, moods, socioeconomic factors or technologies.

Managing an innovation project therefore means finding ways to minimize both the probability of "mistakes"—wrong allocations of resources—and the cost of those mistakes. This means not only looking for the cheapest or fastest path, but also choosing paths that are cheap or quick to adjust and generate valuable learning.

This also implies knowing when a project needs to be adjusted—changing activities—and when it needs to be completely redefined or stopped—changing plans. Maintaining such "double flexibility" or "agility" when managing innovation projects (changing activities or changing plans) requires embedding explicit learning and feedback decision points within the projects.

Hence integrating within the project plan potential "turning points", when significant activities or even the plan as a whole are explicitly reviewed. Turning points can relate in particular to new acquired knowledge and/or to adjustments in the objectives.

An innovation manager is not an astronaut whose module must reach a planet billions of kilometers away based on the optimal configuration of the solar system. An innovation manager is more like a rally driver, moving quickly and taking calculated risks but also constantly receiving and adjusting to new information—"pivoting".

Engage Early

When managing a project with clearly defined work packages and deliverables, it might be adequate to involve its parties over time and only "as needed", providing them with the relevant inputs when it is their turn to act. But "passing the buck" in such a way does not work for sizeable innovation opportunities, when projects are complex, changing and ambiguous, and commitment to evolving objectives is key.

Too often in silo organizations, R&D managers "pass the baton" to the manufacturing department, which then transfers it to sales and marketing, which in turn passes it to purchasing, leading to frustration and inconsistencies. But one of the most important factors that differentiates innovation winners from losers has been identified as the degree of interaction between product or process design and other corporate departments, especially manufacturing and marketing.

The key for managers who deal with innovation opportunities is therefore to find ways to identify and proactively engage early on the internal and external stakeholders who are critical to the launch and scale-up of the opportunity. This generates tangible returns through improved and continuous coordination but also more intangible—though no less important—ones such as increased engagement and motivation.

In corporations with centralized R&D functions, products are sometimes designed by headquarters and then "dumped" to the various customer-facing businesses with the injunction to launch them. The result is too often unbalanced project portfolios, lack of engagement, a longer time to reach the market and management conflicts.

So What?

Managers who deal with innovation opportunities must rethink the way they manage and in particular monitor and review projects. They need to boost winning projects rather than try to save failing ones. They need to be ready to regularly question both the activities—change paths—but also the projects—change maps. They need to focus on future learning and smart experiments rather than on past activities and deliverables. Finally, they need see their role as enablers more than controllers, focused on maintaining momentum, not only alignment.

6.2   Lean Development: Speed and Flexibility

Abandoning traditional decision-making processes and project management approaches when dealing with innovation does not mean switching to chaos and improvisation. It means being able to work in dual modes, balancing effective execution with timely learning and flexibility (Fig. 6.3).

In terms of projects, it also means steering innovation opportunities and teams in a way that combines focused experimentation and ruthless prioritization on the one hand with a supportive environment and the mobilization of people and resources on the other.

Lastly, in terms of resources, it means being able to unleash dedicated capabilities, beyond what can be done just "on top of" day-to-day business activities, both to explore relevant but untapped areas and to exploit potential new business development opportunities.

Key Insights

  i. Capturing innovation opportunities requires crafting *decision-making processes* and working in dual learning modes, with phases—or "stages"—of intensive and focused development and experimentation combined with moments—or "gates"—of questioning and prioritization.
 ii. *Steering innovation projects and teams* requires combining the discipline of focused project portfolio and clear management commitments with an environment that fosters leadership, risk-taking and experimentation.
iii. *Crossing the gap between a fuzzy innovation idea and sizeable value creation* requires dedicated resources and capabilities, both to explore and frame selected potential opportunities and to exploit and scale up potential new businesses. Innovation is real work, requiring significant and specific time and resources.

Fig. 6.3 Learning and execution

6.2.1 Learning While Delivering: Crafting the Right Decision Processes

Capturing innovation opportunities effectively requires combining the efficiency and speed of competitive businesses with the intelligence, flexibility and agility to cope with the uncertainties and ambiguities of innovation-intensive environments. It means going fast while knowing when to change course.

Pacing Commitments and Learning

An effective way to combine speed and intelligence is to replace "one-shot" investment decisions and plans by a sequence of more and more informed but more and more irreversible decisions. The scope and value of the innovation opportunity is then regularly questioned as it unfolds, and commitments are made only as uncertainty and ambiguity decrease.

Venture capitalists have long been known to "stage" their investment decisions along successive financing rounds, where projects are stopped, or are further funded only as the uncertainty surrounding their potential decreases. Some VCs actually focus on specific company development stages, either at the "seed" or "start-up" phases (early stages) or at the expansion, MBO or IPO phases (later stages).

This process will be optimized if at each stage the focus is placed on designing and implementing "cheap experiments", that is, finding the smartest way to test hypotheses and decrease uncertainties as quickly as possible. While many organizations have implicitly or explicitly learned how to "de-risk" their core business projects in such a way, finding out where the greatest uncertainties lie is often a much bigger challenge when focusing on new business development opportunities.

When dealing with new business opportunities, the greatest risks on which the focus should be placed early on might be related to new regulatory, partnership or logistical issues rather than the technical, intellectual property and/or commercial risks industrial companies are used to dealing with. "Sustainability" is an example of such a criterion, which used to be an afterthought in many industries but is now a key issue to consider early on.

Designing Effective Decision Cycles

The best times to stop and question an innovation project before taking a further "go", "no-go" or "freeze" commitment will be a function of the pace, evolution and characteristics of each project and its environment. But when dealing with a large number of innovation projects, it is often more effective to define upfront explicit and common decision cycles, where a portfolio of projects is considered and reviewed. The key here is to define when the decision cycles should happen, who should be involved and how the meetings should be prepared and organized.

Popular ways to implement decision cycles include the "stage-gate" processes (Cooper et al. 2002) known to many organizations as well as the agile management approaches used in software development. In the latter case, work is organized in short cycles with defined successive user-driven goals. During each cycle the focus is on team-driven speed and efficiency, with minimum management interruption.

When Gates Become Hurdles

While the principles of decision cycles that combine speed and intelligence are quite simple, their effective implementation is a never-ending challenge for innovative organizations, in terms of both the actual implementation of the decision-making process and its integration within the corporate governance structure. The decision-making process for innovations must be fine-tuned as a function of the organization's resource and environment, its existing governance and culture and the types of innovation opportunities. There is no one-size-fits-all solution.

Typical decision cycles implemented in various industries will include three to nine cycles ("gates"), implemented over periods of a few months to more than a decade. They include company-specific processes as well as industry standards, such as the four development phases of pharmaceutical drugs or the nine "Technology Readiness Levels" of aeronautical technologies. They can also include locally implemented processes, for example, for incremental process innovations, as well as corporate-wide approaches, such as new business development entities.

Key design issues here include the number of decision-making processes to be implemented in parallel throughout the organization (e.g. one for "local" innovation and another for "strategic" projects) as well as their coordination and sorting. Another key design issue is the permitted level of customization of the processes for specific projects, particularly those involving external partners such as start-ups or academia. Being too rigid vis-à-vis all external partners might be inefficient, but trying to adapt to each of them will often become ineffective.

Finally, governance and administration of the process itself must be carefully defined in terms of metrics and responsibilities, frequency and reporting as well as knowledge and data management procedures and systems. The principles are simple, but their implementation is not.

Staying in Touch with the Core Business

Another important challenge is to keep a consistent link between the decision-making processes related to innovation opportunities and the "normal" management processes of the organization. Key integration issues include the links with "traditional" corporate governance and resource management processes, such as ERPs and annual budget processes; the availability, engagement and timely mobilization of managers—not only when it is too late or very costly to change course—and the risks of corporate cultural bias against "do-check-plan" approaches, with managers reverting to dominant routines and "traditional project review" modes.

Decision cycles can be implemented in various ways, from decision points integrated in the agenda of regular management meetings and supported by small presentations and simple spreadsheets to full-blown dedicated monthly workshops supported by complex software platforms integrating hundreds of projects in parallel. Each organization has to fine-tune and adjust over time what works best for itself.

So What?

Innovation managers who want to effectively capture innovation opportunities must design, implement and integrate in their organizations dedicated but pragmatic and iterative decision-making processes, based on decision cycles that combine speed and efficiency with intelligence and timely questioning.

6.2.2 Discipline and Care: Steering Innovation Projects and Teams

A well-designed decision-making process is useless if it is not maintained and managed effectively on a day-to-day basis. This means maintaining the right balance of discipline and care at project and process levels, combining consistent and sometimes harsh decision-making with the support and attention needed to sustain the engagement of people and teams. A "too disciplined" process will dry up innovation and demobilize people; a too "nice" process will lead to project proliferation, wasted resources and delayed implementation.

Staying on Course

A key success factor when steering innovation projects, particularly in the early stages, is maintaining focus on reducing uncertainties quickly and intelligently. This means that the project should be planned and continuously managed in order to maximize its "learning over investment".

Fast-prototyping, beta-testing, lead-users and crowdsourcing are examples of innovation management techniques that can be used to learn in a faster and cheaper way than when using traditional product or process development approaches.

A second key success factor is maintaining discipline and clarity even as a project repeatedly changes course. This especially entails ensuring that the project ownership and scope remains consistent and clear, and that flexibility does not become an excuse for ambiguity. This also means that the decisions taken regarding the project—go, no-go, try again or freeze—are explicitly made, agreed on and communicated. Lastly, it means that the implications in terms of corporate resource allocations are acknowledged and dealt with by the organization and its management.

This second key success factor is particularly critical in the latest stages, when the innovation project is handed over to—or sometimes "dumped" on—the "normal" organization for implementation, scale-up and/or commercialization. Too often this implementation step falls in the gap between the teams focused on innovation, for which the project is perceived as "ready to launch", and the teams focused on day-to-day business activities, for which it is not mature enough and/or not yet a priority.

The curse of innovation managers is the "zombie project", which is vague, buzzword-filled and "sexy" enough to prevent anyone from taking the responsibility to stop it, yet so risky, complex or "nice-to-have" that nobody is actually ready to commit any significant resources to support its development.

Sustaining a Balanced Portfolio

As existing innovation projects fail, struggle or are redirected, nurturing and maintaining a consistent and focused portfolio also requires a continuous inflow of new projects and of new input and energy in existing projects.

The differentiating factors between successful and unsuccessful venture capitalists are mainly related to both the quality of their deal flow (the new projects they constantly receive) and the focused drive and support they bring to the ventures in which they have invested.

This means first that the portfolio of innovation projects being implemented must be carefully monitored, and that dedicated initiatives aimed at "refilling" the pipeline of innovation opportunities must be regularly launched. The initiatives should leverage both internal and external sources and harvest both early-stage ideas and more mature emerging opportunities.

Second, it means that the involvement and commitment of the project team members, process sponsors and key resource controllers must be sustained, even under the short-term pressures of the day-to-day business. It is particularly critical to maintain a high level of leadership, risk-taking and experimentation among project owners and sponsors, even as many projects end up being stopped and resources are reallocated. Without proper incentives and support, a continual start/stop mode can demotivate or even over-stress the people involved.

One of the biggest challenges for innovation champions is to accept that their project can be stopped even if they delivered what they had promised, and that they might be reallocated to a new project that also require their full energy, attention and motivation. For their sponsors, the challenge is to maintain focus and attention while coping with the pressure and emergencies of their day-to-day activities.

So What?

Innovation managers who design a dedicated decision-making process for innovations must also ensure that dedicated "process owners" and sponsors are capable and in charge of maintaining the discipline and momentum required for the processes to be effective. What matters is to combine excellence and agility, not bureaucracy and compliance. Processes that exist only "on paper" only generate more paper.

6.2.3 From Fuzzy Front End to Value Creation: Crossing the Gap

The implementation of an innovation opportunity is particularly challenging when what is at stake is the whole development of a new business, mobilizing significant corporate resources through new technologies, toward a new target market and/or a new value chain.

The journey from an innovation idea to a sizeable new business is long and tortuous, and too often organizations believe that it can be easily completed, in a few years and with marginal resources. But most sizeable innovative businesses, even digital ones, require significant resources and effort to launch and will take five to ten years to become significantly profitable.

Too often executives hope that ideas harvested from employees can be simply turned into sizeable corporate profits in 36 months. But even legendary successes such the Post-it or Facebook did not achieve that. And a quick look at their own corporate history often shows that past sizeable corporate successes took many years to flourish and required millions in investment.

Taming the Fuzzy Front End

Identifying new business development opportunities in an effective way does not mean simply unleashing employees' creativity in order to generate multiple "out of the box" ideas "on top of" their existing jobs. It needs to be done in a systematic way, and with dedicated resources.

The first issue is to clarify as much as possible the "new box", that is, what the firm considers its legitimate "strategic playground", based on its purpose, resources and environment. This means defining the type of new activities it could potentially engage in, as well as those it currently considers beyond the firm's scope.

A pharmaceutical firm focused on malaria might consider as part of its "strategic playground" the development of a new vaccine or treatment but might exclude as outside the scope of its core business other types of eradication approaches, such as developing new water treatment processes, mosquito nets or promising technologies aimed at modifying the DNA of entire mosquito species. If you do not know what you are looking for, you are unlikely to find it.

The second issue is to identify and prioritize untapped areas within the strategic playground, where the firm wants to focus its new business development effort. This step must involve the firm's key stakeholders, as it will define where it is perceived as legitimate to explore, and where it is not. A failure to do so will increase the risks of the corporate immune system's rejecting the future new business transplant as a "not invented here" alien.

An industrial firm might define its "strategic playground" around its core technologies but also identify as legitimate but untapped areas digitization and/or servitization opportunities. In a similar way, a B2B firm might identify opportunities to move downstream in the value chain as untapped but potentially relevant. As Michael Porter famously said, "Strategy is about deciding what not to do".

The final issue is to set up dedicated processes to explore untapped but legitimate areas and to identify specific business development opportunities. This can be done by combining bottom-up approaches based on open challenges and online communities with more top-down approaches aimed at systematically screening specific emerging technologies, markets and/or industries. The key here is to effectively engage both internal and external resources, as the skills and knowledge needed to explore the untapped areas is likely outside the corporation's comfort zones.

Such targeted search and screen processes might include dedicated events, such as corporate hackathons, and the systematic due diligence of a large number of potential partners and/or acquisition targets. Heaven helps those who help themselves.

A common mistake here is to focus only on new "greenfield" ideas, which can be very risky and time-consuming to develop. It is often more profitable for a corporation to acquire a small or medium-sized business and scale it up, rather than to start from "scratch".

Growing Up

The identification of a potential new business development opportunity—the exploration—is actually the easy part of the process. The key challenge is to turn an emerging or small venture into a profitable corporate entity—the exploitation—rather than a mere footnote in the corporate annual report.

Most climbing fatalities on dangerous mountains occur during the descent. Developing a sizeable and profitable new business once an attractive opportunity has been identified ("descending after the summit has been reached") is one of the toughest jobs of innovation.

The first issue is to gain an in-depth understanding of the main value drivers and key performance indicators of the new business, which will often differ from those of the parent corporation. Failure to do so can lead many corporations to under- or overvalue specific opportunities and/or develop them in unsustainable way.

A B2B business focused on return on investment and utilization rates might mismanage a new online B2C business where average revenue per user and churn matter more. An automotive manufacturer focused on supply chain management and cost control will have different key performance indicators than a data-driven electric car or mobility platform developer.

The second issue is to set up a clear, flexible and effective governance process for the new business implementation, which is close enough to the parent corporation to capture synergies but also autonomous enough to maintain momentum and flexibility. This means in particular allowing the new business to access across the corporate silos and in a flexible way the critical resources it needs to fuel its growth, such as talent, cash, technologies or facilities. It also means ensuring that strategic investment decisions can still be made in a "fail early and succeed fast" mode, shaking free of traditional budgeting processes, risk aversion and resistance to change. Finally, it means clarifying and managing the involvement of the top management and of key talents between both the core business—often pressing—and the new business—often quickly changing—priorities.

A typical new business venture might be managed by a new business board that reports to the CEO or the corporate board, and includes both experienced insiders and more entrepreneurial or diverse outsiders. Staffing that board and the new venture with the right people and maintaining their engagement will often be more critical than the quality of the underlying technology.

So What?

Sizeable and profitable new businesses do not "pop up" out of the blue and "on top of" day-to-day business activities. They require dedicated resources and systematic processes to identify and screen untapped but legitimate opportunities and then scale them up while leveraging synergies with the parent corporation.

6.3   Smart Money: Funding Innovation Projects

Innovation projects often require significant and sustained upfront investments before achieving profitability. But traditional funding approaches,
based on predefined targets and corporate budgets or bank loans, are in many
cases ill-suited to the uncertain and changing world of innovation.

Innovation managers must therefore understand the various alternative
funding options available, as well as the peculiarities of the specialized investors they can call on (Fig. 6.4).

Key Insights

i.   The lack of track record, the ambiguity and the specificity of most innovation
projects mean that innovation managers should consider alternative specialized *sources of financing*, such as venture capital, to (co-)invest in the development of their opportunity.

ii. Innovation managers and entrepreneurs looking for specialized outside
(co-)investors must identify, target and engage effectively the right type of
financial stakeholders, based on available offers, expectations and potential
value-added.

iii. Innovation managers and entrepreneurs can in some cases leverage the
value-added and support of *venture capitalists* as (co-)investors, provided
they master the business model of these specialized investors, their selection
criteria and their negotiation process.

Fig. 6.4   Smart money

6.3.1 Sources of Financing: Why Cash Is Costly

While some "frugal" or "incremental" innovation projects can be developed with limited means and available sources of cash, most radical innovation and new business development opportunities cannot be captured without significant upfront investment. In order to be able to mobilize the financial resources needed to develop these projects, innovation managers and entrepreneurs must understand their peculiarities from the perspective of a potential investor as well as the implications in terms of available funding options.

Innovation managers or entrepreneurs are sometimes frustrated because they feel that "finance people don't get it". They forget that their role as innovators is to identify, understand and convince key stakeholders, including investors. Would they lend their own savings account to an innovator?

Apple or Lemon?

The first challenge for investors and thus for the innovators trying to convince them is that they often have trouble understanding and assessing what exactly they are investing in. The first source of this "information asymmetry" is the inherently complex and ambiguous nature of innovation projects, leading to potential adverse selection. This is particularly the case when investors cannot rely on personal trust in the innovator and/or its team.

The second source of information asymmetry is the typically limited history as a business activity of innovation projects, which therefore cannot provide the formal financial reporting and indicators most investors are used to dealing with. This information asymmetry also means that the cash from operations provided by traditional budgeting processes is often more difficult to mobilize for innovation projects.

Some investors cope with the newness, complexity and ambiguity of innovation projects by specializing in specific areas, such as biotechnologies or advanced materials, and by relying significantly on the track record of the innovator and its team rather than on the investment proposal itself.

Handle with Care

The second challenge for investors and thus for the innovators trying to convince them is that innovation investments are more complex to handle and less "liquid" than traditional investments, creating in particular significant "agency costs" (Jensen and Meckling 1976).

The first cause of these "agency costs" is the need to assess, monitor and manage innovation investments, which often represents as a cost a non-negligible share of the invested amounts. This is particularly the case when considering the incomplete contracts and significant moral hazard often involved in dealing with innovation projects.

Managing a €500,000 stake in a venture requires in most cases nearly as much effort and attention as managing a €50 million stake, while the upside potential of the former is obviously much smaller. This means that innovators must struggle to make their opportunities big enough to justify the fixed costs of investors' attention.

The second cause of these agency costs is that any investor who wants to exit or sell an innovation investment might find himself left with very limited tradable assets. Moreover, the assets can be very specific, that is, worthless or of limited value for buyers other than the innovator. On top of this, the R&D activity often involved might be difficult to scale down—one cannot sell "a slice of it"—and cannot in most cases be capitalized as an investment. Such agency costs are why the loans provided by traditional financial institutions or large suppliers are often more difficult to mobilize for innovation projects.

Venture capital investments in innovation projects might in some cases generate high returns but tend to involve significant management fees and very limited liquidity. Public investments in specific technologies ("picking winners") might also end up being completely worthless sunk costs if the technology fails to deliver the expected benefits.

Find Your Rich Uncle

The information asymmetries and agency costs involved in innovation projects mean that traditional funding sources such as corporate budgets and bank loans are often neither suitable nor accessible, or can provide only a small share of the full financing. Innovation managers must therefore consider alternative funding sources as investors or co-investors, particularly when dealing with radical innovation and new business development opportunities.

Alternative funding sources for innovation managers include their own management team and other "informal" individual investors—for small-scale or early-stage projects—as well as professional investors such as equity funds, the corporate venture capital arms of existing firms as well as regional or thematic providers of public subsidies and grants.

So What?

Innovation managers who try to convince potential investors to fund the upfront investments required by their project must first identify, understand and minimize the information asymmetries and agency costs that are intrinsic to their project, then find and target the specialized investors who are ready to cope with these hurdles.

6.3.2 Navigating the Innovation Financing Fauna: Financial Stakeholders

Investors in an innovation project are often key stakeholders in its development. Innovation managers must therefore carefully select and manage them, considering both those most likely to "buy" the project as well as those most likely to add value and significantly contribute to its success.

Dumb Money

The simplest financial stakeholders to deal with are probably the traditional financial institutions, whose business model is to lend money at a guaranteed rate and with the protection of predefined collaterals. They indeed require limited attention from innovation managers once the loan terms have been negotiated. Conversely, they also in most cases provide limited value-added to the project, beyond somehow signaling its credibility.

More "sophisticated" lenders include specialized suppliers or lead customers, ready to pre-finance a project, for example, in exchange for temporary exclusivity. While these sources of funding are probably the first ones to pursue, they are in many cases not sufficient when dealing with innovation projects, given the limited collaterals available in most cases.

Bank loans and other traditional sources of debt such as credit cards are actually by far the first source of funding for new businesses. But their limited scope and rigidity make them unsuitable for most truly innovative ventures.

Smart Mothers-in-Law

Innovation managers or entrepreneurs with limited collateral and/or who look for less risk-averse financial stakeholders can call on a wide range of informal and professional equity investors, who take a share of the risks in exchange for a share of the future benefits. In most cases, they also get directly or indirectly involved in the governance of the innovation project. (Corporate) entrepreneurs who strive for autonomy must therefore be ready to give up some power and control, which is not easy for many of them.

The key challenge for innovation managers is therefore not limited to convincing potential investors to risk their fund; they must also design and manage an appropriate governance structure to involve them in an effective way. Given the—actual or perceived—loss of control this involves, finding "smart" investors who will actually add value to the venture is critical, be it through their skills, reputation, experience or networks.

Some "unicorn" start-ups have managed to raise significant amounts of money from investors while allowing the founders to retain most of the control. But in many cases young start-ups are so desperate to close their initial funding rounds that they end up with a large and eclectic set of investors. In the worst cases, this will consume a lot of management time, create conflicts and hinder future growth.

Finding Wisdom and Crowds

The first set of equity investors which (corporate) entrepreneurs can call on is the informal investors, individuals who invest their own money. They are particularly relevant for small-scale and/or early-stage projects, where small amounts, personal involvement and limited due diligence are involved.

Informal investors include the management team and their relatives (the "friends, fools and family"; Kotha and George 2012), individual investors not directly related to the management teams ("business angels") and the wider public (through crowdfunding; Belleflamme et al. 2014). Again, here the challenge is to find and work with "smart" money, that is, find investors with the potential to add value to the ventures beyond their financial investment, for example, by providing expertise, feedback or visibility.

For centuries people have reached out to the public in order to fund risky ventures, for example, to fund the building of the Statue of Liberty. The wide availability and reach of internet-based technologies have simplified and popularized such approaches but have not solved the underlying governance issues. Then and now, it can lead to nice popular initiatives but also, in the worst cases, to serious fraud and scams.

Business angels, that is, individual investors who invest their own money without having personal connections to the corporate entrepreneur or its team, can in particular play a positive role in the development of an innovative venture. Their informal approach to innovation and their personal engagement in the venture can provide valuable coaching and support to the (corporate) entrepreneur and his or her team, even if these "gray-haired" investors sometimes overestimate the relevance of their skills or experience.

Business angels (Maxwell et al. 2011) are typically high net-worth individuals who invest their own money directly in a small number of local ventures. They expect financial gains from these risky investments but also personal satisfaction and "fun" from their hands-on interaction with the venture and its team. Their personal affinity with the project and/or team can therefore play a significant role.

Working with the Pros

When informal investors are unavailable or inadequate, the innovation manager or (corporate) entrepreneur can also call on "professional" investors who manage funds entrusted to them by others. They include on the one hand high-risk/high-potential venture capitalist funds (De Clercq et al. 2006) and on the other hand public and not-for-profit funds. These investors will often rely on extensive due diligence and detailed contracts. They will also tend to carefully monitor and control the venture's development and impact, be it in terms of growth or visibility. Finally, they will often focus only on high-impact projects related to specific themes and/or regions.

While professional investors are often more sophisticated than informal investors, they are also more demanding in terms of due diligence, governance and reporting. They are also in general more restrictive in terms of the type of ventures they consider and the venture's "fit" with their target scope. (Corporate) entrepreneurs must therefore make sure that they target the "right" investors, given the characteristics of their venture, and that the cost/benefit trade-off of involving them is positive.

Professional public and private investors have played a major role in the development of "general purpose" technologies such as telecommunication and computing technologies, biotechnologies and advanced materials.

So What?

In the same way that they need to know their customers and understand their needs, innovation managers and (corporate) entrepreneurs must know the financial stakeholders they face and understand their respective needs and expectations. They also need to focus on "smart" money and design governance structures that allow their investors to effectively share control and add value.

6.3.3 Professional Innovation Investors: Venture Capitalists

Innovation managers who deal with high-risk/high-potential opportunities should consider working with specialized professional (co-)investors. These "venture capitalists" can provide direct financial and managerial support to individual ventures but also indirect learning in terms of deal-flow management, due diligence, deal-making, growth acceleration and exits. However, innovation managers have to understand how venture capitalists work and how they pick, negotiate and manage investments.

Financial Bimbos

The most important characteristic of venture capitalists and of how they differ from corporations is that they "marry in order to divorce". The business model of most venture capitalists is to sell—"exit"—the ventures they have invested in, as quickly as possible and with the highest return.

Venture capitalists raise cash from investors and then try to invest the cash by buying stakes in high-growth ventures, which they will then be able to "flip" by selling them, in most cases to corporations through trade sales or to public markets through an initial public offering ("IPO"). The focus when they invest is therefore on how to attract the most "sellable" ventures and on when and how to profitably "exit" them.

As a consequence, the best venture capitalists are able to attract a deal flow with the most attractive ventures, and to ruthlessly stop funding laggards in order to support, accelerate the growth of and expedite profitable exits from the most valuable ones. In particular, their ability to identify and regularly assess the growth potential of existing ventures and teams, and to aggressively scale up the most promising ones, generates significant opportunities for the corporations with whom they work to learn and share good practices.

Corporate venture capital (CVC) approaches are one example of a potential approach for large corporations to work with and learn from venture capitalists, provided they can manage the time horizons and risks involved, as well as the necessary trade-offs between the strategic fit and financial returns of potential ventures.

Swimming with Sharks

One key success factor for venture capitalists is their ability to squeeze as much value as possible from the successful ventures they have invested in while minimizing their risks, costs and exposure. As a consequence, they have developed dedicated skills regarding how to work with other investors—through "syndication"—and how to negotiate and close attractive deals—valuation.

Venture capitalists have long practiced open innovation, by jointly investing in innovation ventures. Such "syndication" (Manigart et al. 2006) allows them to minimize their exposure to individual ventures and to maximize liquidity by diversifying their investments. It also allows them to improve their deal flows by extending their reach, reputation or networks. Lastly, it allows them to capture synergies with other investors through resource and information flow-sharing, expertise cross-fertilization and increased bargaining power.

Many venture capital investments will include a "lead" investor who takes charge of the deal's negotiation and management, supported by a small number of "co-investors" who share the ride. In these cases, a key challenge for innovation managers and entrepreneurs is to manage over time the sometimes diverging priorities and expectations of the various investors.

Venture capitalists have also developed unique expertise in terms of deal negotiation and valuation in complex and risky environments. They understand that the value of an innovation opportunity is driven by "objective" factors such as discounted cash flows, team quality and competitive advantage, but also by more subjective ones such as relative bargaining power and negotiation skills, risk aversion, potential synergies between the investors and the venture as well as more general trends, cognitive models and moods—what is "hot" and what is not.

A typical venture capitalist will invest in less than 1% of the high-growth ventures it considers, will be able to cope with failure rates above 50% and will exit the successful ventures after on average five to seven years and with an expected ten- to twentyfold return. Only a small fraction of the investments will reach their target but those few "winners" will generate the bulk of the profits.

So What?

Innovation managers involved in radical innovation and new business development opportunities should work with and learn from professional innovation investors, such as venture capitalists. Best practice sharing opportunities include deal-flow generation, valuation and prioritization as well as syndication and growth acceleration.

6.4   Synthesis

Fail Fast and Win Big: Key Insights

6.1. Nimble execution: learn cheaply and adapt quickly

i. Management is (maybe) a science, but *innovation is an art*. Traditional corporate decision-making and project management approaches are therefore ill-suited to the ambiguity, high failure rate, pace and multifunctional aspects of innovation.

ii. *Innovators must play poker* rather than chess. Traditional corporate decision-making approaches ("chess") rely mainly on analyzing facts and minimizing failures—"thinking first". But managing innovative organizations and capturing sizeable innovation opportunities ("poker") must also rely on experimentation and proactive learning, combined with ruthless prioritization—"doing first".

iii. While traditional project management approaches rely mainly on set targets and task allocations, managing innovative organizations and capturing innovation opportunities also implies *planning for changes in the plan* and focusing on embedded flexibility and cross-functional mobilization.

6.2. Lean development: speed and flexibility

i. Capturing innovation opportunities requires crafting *decision-making processes* and working in dual learning modes, with phases—or "stages"—of intensive and focused development and experimentation combined with moments—or "gates"—of questioning and prioritization.

ii. *Steering innovation projects and teams* requires combining the discipline of focused project portfolio and clear management commitments with an environment that fosters leadership, risk-taking and experimentation.

iii. *Crossing the gap between a fuzzy innovation idea and sizeable value creation* requires dedicated resources and capabilities, both to explore and frame selected potential opportunities and to exploit and scale up potential new businesses. Innovation is real work, requiring significant and specific time and resources.

6.3. Smart money: funding innovation projects

i. The lack of track record, the ambiguity and the specificity of most innovation projects mean that innovation managers should consider alternative specialized *sources of financing*, such as venture capital, to (co-)invest in the development of their opportunity.

ii. Innovation managers and entrepreneurs looking for specialized outside (co-)investors must identify, target and engage effectively the right type of *financial stakeholders*, based on available offers, expectations and potential value-added.

iii. Innovation managers and entrepreneurs can in some cases leverage the value-added and support of *venture capitalists* as (co-)investors, provided they master the business model of these specialized investors, their selection criteria and their negotiation process.

Bibliography[1]

Belleflamme, P., Lambert, T., & Schwienbacher, A. (2014). Crowdfunding: Tapping the right crowd. *Journal of Business Venturing, 29*(5), 585–609.

Cooper, R. G., Edgett, S. J., & Kleinschmidt, E. J. (2002). Optimizing the stage-gate process: What best-practice companies do—I. *Research-Technology Management, 45*(5), 21–27.

De Clercq, D., Fried, V. H., Lehtonen, O., & Sapienza, H. J. (2006). An entrepreneur's guide to the venture capital galaxy. *The Academy of Management Perspectives, 20*(3), 90–112.

Jensen, M. C., & Meckling, W. H. (1976). Theory of the firm: Managerial behavior, agency costs and ownership structure. *Journal of Financial Economics, 3*(4), 305–360.

Kotha, R., & George, G. (2012). Friends, family, or fools: Entrepreneur experience and its implications for equity distribution and resource mobilization. *Journal of Business Venturing, 27*(5), 525–543.

Manigart, S., Lockett, A., Meuleman, M., Wright, M., Landström, H., Bruining, H., & Hommel, U. (2006). Venture capitalists' decision to syndicate. *Entrepreneurship Theory and Practice, 30*(2), 131–153.

Maxwell, A. L., Jeffrey, S. A., & Lévesque, M. (2011). Business angel early stage decision making. *Journal of Business Venturing, 26*(2), 212–225.

Mintzberg, H., & Waters, J. A. (1985). Of strategy: Deliberate and emergent. *Strategic Management Journal, 6*(3), 257–272.

Wiltbank, R., Dew, N., Read, S., & Sarasvathy, S. D. (2006). What to do next? The case for non-predictive strategy. *Strategic Management Journal, 27*(10), 981–998.

---

[1] An extended bibliography is available at www.NavigatingInnovation.org

7

Conclusion: More Brain, Less Storming

Managing innovations requires understanding what innovation means, why an organization needs and wants to innovate and the capabilities it should develop as a consequence. This might imply that your organization should implement the latest fad or emulate Apple and Uber. Or that it should not.

Managing innovations means thinking about how to build an organization that is continuously able to identify, assess and implement new opportunities in line with its strategic objectives. This means in particular being able to steer people and resources in a changing world, where knowledge and objectives are not a priori given but are framed and adjusted over time.

There is no universal recipe for successful innovations. Each organization has to identify and prioritize its needs based on its purpose, resources and environment. What was right for Google, 3M or Tesla might not be right for your organization here and now.

Managing innovations also entails mobilizing stakeholders and implementing cheap and iterative learning experiments, in order to cope with uncertainties and resistance to change. Managing innovations implies turning your and other people's new ideas into new realities. This might involve fostering more creativity and ideation, or might not. Inventors are convinced, but innovators are convincing.

Managing innovations requires a bit of storming and a lot of brains.

Innovations have allowed billions of people to live longer and better lives. They have also allowed humans to land on the moon and occupy nearly every corner of our planet. But innovations also threaten cultures, communities and ecosystems. They disrupt people, whose values, skills and assets fall into obsolescence. They destroy social institutions as well as fragile and unique environments. They create winners but also losers.

© The Author(s) 2018
B. Gailly, *Navigating Innovation*, https://doi.org/10.1007/978-3-319-77191-5_7

Managing innovations therefore also requires finding ways to tame innovations and their consequences. It requires managers to think and choose what should be done and what should not be done. It means mastering the dark side of innovation, in order to create a planet where every present and future human being can find his or her place. A fool with a tool is still a fool. We develop great tools; let us not become great fools.

Managing innovations requires a soul.

Index

© The Author(s) 2018
B. Gailly, *Navigating Innovation*, https://doi.org/10.1007/978-3-319-77191-5